Basic Electrical Power and Machines

Basic Electrical Power and Machines

David Bradley
Engineering Department
School of Engineering, Computing and Mathematics
Lancaster University, UK

CHAPMAN & HALL
University and Professional Division
London · New York · Tokyo · Melbourne · Madras

Published by
Chapman & Hall, 2–6 Boundary Row, London SE1 8HN, UK

Chapman & Hall, 2–6 Boundary Row, London SE1 8HN, UK

Blackie Academic & Professional, Wester Cleddens Road, Bishopbriggs, Glasgow G64 2NZ, UK

Chapman & Hall Inc., One Penn Plaza, 41st Floor, New York NY 10119, USA

Chapman & Hall Japan, Thomson Publishing Japan, Hirakawacho Nemoto Building, 6F, 1-7-11 Hirakawa-cho, Chiyoda-ku, Tokyo 102, Japan

Chapman & Hall Australia, Thomas Nelson Australia, 102 Dodds Street, South Melbourne, Victoria 3205, Australia

Chapman & Hall India, R. Seshadri, 32 Second Main Road, CIT East, Madras 600 035, India

---

First edition 1994

Typeset in 10/12 Times by Cotswold Typesetting Ltd, Gloucester, England
Printed in Great Britain by Alden Press, Oxford

ISBN 0 412 45540 4

A catalogue record for this book is available from the British Library

Library of Congress Cataloging-in-Publication data

Basic electrical power and machines/David Bradley. – 1st American ed.
p. cm.
Includes index.
ISBN 0–412–45540–4
1. Electric machinery. 2. Power electronics. 3. Mechatronics. 4. Electric transformers. I. Title.
TK2182.B73 1994
621.31′042–dc20 93-34700
CIP

♾ Printed on acid-free text paper, manufactured in accordance with ANSI/NISO Z39.48-1992 (Permanence of Paper).

Contents

Preface

Electrical machines range in size from the turbo alternators used in power stations and developing hundreds of megawatts to micromachines fabricated in silicon and capable of delivering only a few nanowatts. In the form of a.c. induction motors and d.c. machines they provide the bulk of the drives used in industry and the home while more specialist machines such as servomotors and stepper motors are used for applications requiring precise and controlled positioning. The influence of electrical machines and drives therefore extends across the whole of society and their characteristics and use have relevance not only to electrical and electronic engineers but also to mechanical, and indeed civil, engineers.

The development of power electronics technology has enabled the introduction of advanced control systems and strategies and the shaping of machine shaft characteristics. This has led to the integration of electrical machines and drives into a wide variety of systems from robots to cameras and machine tools to washing machines. Indeed, it is the development of advanced intelligent machine controllers that lies at the heart of the development of the concept of mechatronics which is integral to the design of many such systems.

The material in the text is based upon courses on electrical machines, power electronics and mechatronics which have been taught to electrical, electronic and mechanical engineers at Lancaster University over a number of years. It is not intended as a text for designers of electrical machines but is intended as an introductory text to the subject with the emphasis placed on machine operation and the control and modification of the terminal and shaft characteristics of a range of electrical machine types.

However, though the emphasis is on electrical machines, it must not be forgotten that there are available a range of pneumatic and hydraulic drives which have applications in areas where electrical machines of the required characteristics and performance are not available. A final chapter setting out the basic operation of such machines is therefore included for comparison purposes and to make the reader aware of the range of alternatives.

Finally, the writing of a text such as this has an impact not only on the author but also to a greater or lesser degree on a number of other individuals, and thanks must therefore be given to all those who have contributed in some way to its preparation. First, and far too numerous to mention, are those students who have taken the various courses referred to above and whose responses – solicited and otherwise – have all contributed to the development of those courses and hence to this text. Second, there are those of my colleagues in the Engineering Department at Lancaster University on whose ordered existence the preparation of material somehow impinged. In this category mention must be made of David Dawson who provided much of the base material on hydraulic and pneumatic drives and who, along with Derek Seward, patiently provided support in other ways, my particular thanks go to them. Finally, my thanks go to Bob Morrison for taking the trouble to read and comment on the first full draft of the text and for his helpful comments.

List of symbols

A	Cross-sectional area
A	Amperes
B	Magnetic flux density (tesla (T) or weber per metre (Wb/m))
C	Capacitance
D	The specific displacement in m^3/radian of a hydraulic motor
$\boldsymbol{E}$	Electromotive force (e.m.f.)
e	Instantaneous value of e.m.f.
F	Force (newtons (N))
$\boldsymbol{F}$	Magnetomotive force (m.m.f.) (ampere turns/m)
f	Frequency (Hz)
$\boldsymbol{H}$	Magnetic field (A/m)
J	Inertia (N m s^2), inertia of a rotating system
K	General constant term
L	Inductance
$\boldsymbol{l}$	Vector length
M	Mutual inductance
N	Number of turns
n	Rotational speed in rev/min
P	Power (watts)
p	Number of poles of an electric machine or supply pressure in a hydraulic system
p	Converter pulse number or pole pairs of a machine
Q	Reactive power (VAr) or volume flow rate (m^3/s)
R	Resistance, thermal resistance or reluctance
$\boldsymbol{S}$	Rotating vector or volt-amperes
T	Torque (N/m)
t	Time
v, i, s, etc.	Instantaneous value of a time dependent signal
$\hat{V}$, $\hat{I}$, $\hat{S}$, etc.	Maximum value of a time dependent signal or vector
V, I, etc.	r.m.s. value of a time dependent voltage or current or the value of a d.c. voltage or current
$\hat{\boldsymbol{V}}$, $\hat{\boldsymbol{I}}$, etc.	Phasor using the maximum value of a voltage or current
$\boldsymbol{V}$, $\boldsymbol{I}$, etc.	Phasor using the r.m.s. value of a voltage or current
$\boldsymbol{V}^*$, $\boldsymbol{I}^*$, etc.	Complex conjugate of a voltage or current phasor
W	System energy component
X	Reactance
$\boldsymbol{Z}$	Complex impedance
Z	Magnitude of a complex impedance
$Z(t)$	Transient thermal impedance of a semiconductor device
α	Thyristor firing delay angle (firing angle)
β	Firing advance angle of a converter operating in inverting mode
δ	Power angle of a synchronous machine
ϕ	Phase angle
$\boldsymbol{\Phi}$	Magnetic flux (weber)

λ	Effective flux linkages (magnetic flux × number of turns)
μ_o	Permeability of free space ($4\pi \times 10^{-7}$)
μ_r	Relative permeability
μ	Current distortion factor
θ	Temperature or the swash plate angle of a hydraulic motor
ρ	Resistivity (ohm, metres)
τ	System time constant
ω	System frequency ($2\pi f$) in radians per second, also synchronous speed of an a.c. machine expressed as electrical radians per second in terms of the supply frequency
ω_{rm}	For a machine, the angular velocity of the shaft in mechanical radians per second
ω_{sm}	For a machine, the synchronous speed expressed in mechanical radians per second (For a p pole machine $\omega = p\omega_{sm}/2 = p\omega_{sm}$)
ω_m	For an induction machine the actual speed of the rotor in mechanical radians per second
ω_{me}	For an induction machine the actual speed of the rotor in electrical radians per second

Alternating current systems 1

Alternating current (a.c.) systems offer a number of advantages over **direct current** (d.c.) systems for the transmission of electrical energy, in particular the ability to readily change the magnitude of the operating voltage by means of transformers, and have come to predominate the transmission and distribution of electrical power.

In a typical a.c. system a source voltage (v) varying sinusoidally with time in the manner of Fig. 1.1 drives a sinusoidally varying current (i) through the connected circuit. This current will also be sinusoidal in form and will have a relationship with the source voltage which is dependent on the nature of the impedance of the connected circuit.

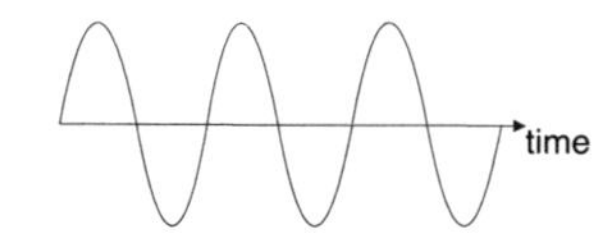

Fig. 1.1 Sinusoidally varying waveform.

1.1 Phasors

Phasors provide a means of expressing information on the **amplitude** and **relative phase** relationships of a group of sinusoidal signals at the same frequency in a form which is independent of time. The use of phasors therefore implies a **steady state** system of sinusoidal signals in which the zero reference for time is entirely arbitrary. In practice, one signal is usually taken as reference and the time relationships of all the other signals are then expressed relative to this reference.

Figure 1.2 shows the generation of a sinusoidal waveform from the projection of a **rotating vector** ($\boldsymbol{S}_1$) of magnitude $\hat{S}_1$ rotating at an **angular velocity** ω on to the **imaginary axis** in the **complex plane**. The instantaneous position of the $\boldsymbol{S}_1$ vector in the complex plane is defined in terms of its angular velocity and its **angular position** (α) at time $t=0$. Using the complex form of equation (1.1) in which $e^{j\omega t}$ represents a **unit vector** rotating at an angular velocity ω.

$$\boldsymbol{S}_1 = \hat{S}_1 \,.\, e^{j\alpha} \,.\, e^{j\omega t} \tag{1.1}$$

Hence the associated sinewave (s_1) has the form:

$$s_1 = \hat{S}_1 \,.\, \sin(\omega t + \alpha) = \mathcal{I}m(\boldsymbol{S}_1) = \mathcal{I}m(\hat{S}_1 \,.\, e^{j\alpha} \,.\, e^{j\omega t}) \tag{1.2}$$

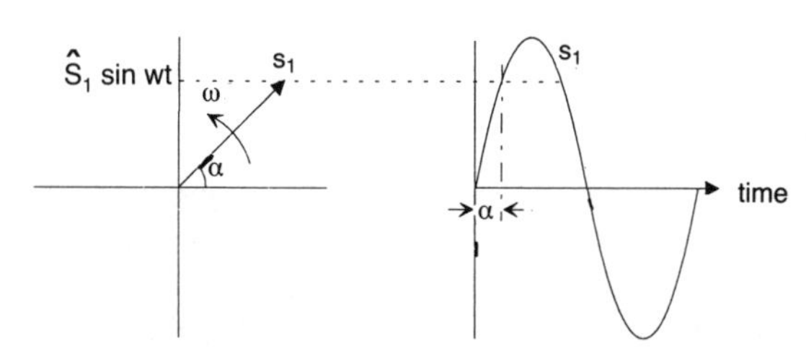

Fig. 1.2 The generation of a sine wave by a rotating vector.

Comment

$\mathscr{I}m(\boldsymbol{S}_1)$ is the imaginary part of vector $\boldsymbol{S}_1$, since

$$e^{j\alpha}e^{j\omega t}=e^{j(\omega t+\alpha)}=\cos(\omega t+\alpha)+j\sin(\omega t+\alpha)$$

Consider now the introduction of a second vector ($\boldsymbol{S}_2$) of magnitude $\hat{S}_2$ rotating at the same angular velocity ω but with a different amplitude and an angular position at time $t=0$ of β. The instantaneous position of this second vector in the complex plane is represented by

$$S_2=\hat{S}_2\,.\,\sin(\omega t+\beta)=\hat{S}_2\,.\,e^{j\beta}\,.\,e^{j\omega t} \tag{1.3}$$

The relationship between the instantaneous values of the sinewaves s_1 and s_2 would then be as indicated in Fig. 1.3 on which $\phi=\alpha-\beta$.

In an a.c. system currents and voltages vary **sinusoidally** with time at a **frequency** determined by the supply frequency and with phase relationships which are a function of the system **impedances.** For example, in Fig. 1.3, signal s_1 can be said to **lead** signal s_2 by the angle ϕ or, alternatively, s_2 could be said to **lag** s_1 by the angle ϕ.

Comment

The phase angle of a signal defines the time relationship between the signal and the reference signal.

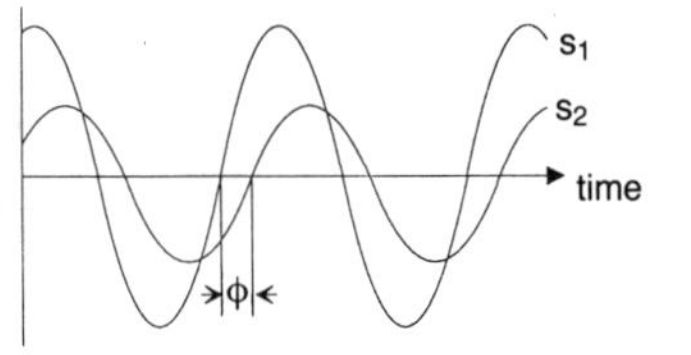

Fig. 1.3 The relationship with time between a pair of sinewaves.

Referring to Fig. 1.3 and choosing s_1 as reference, the s_1 and s_2 sinewaves can be written using the complex form of the expressions for the $\boldsymbol{S}_1$ and $\boldsymbol{S}_2$ vectors as:

$$s_1=\hat{S}_1\,.\,\sin\omega t=\mathscr{I}m(\boldsymbol{S}_1)=\mathscr{I}m(\hat{S}_1\,.\,e^{j\omega t}) \tag{1.4}$$

and

$$s_2=\hat{S}_2\,.\,\sin(\omega t-\phi)=\mathscr{I}m(\boldsymbol{S}_2)=\mathscr{I}m(\hat{S}_2\,.\,e^{-j\phi}\,.\,e^{j\omega t}) \tag{1.5}$$

in which the negative phase angle in the expression for s_2 indicates that s_2 is lagging s_1.

Alternatively, choosing s_2 as reference:

$$s_1=\hat{S}_1\,.\,\sin(\omega t+\phi)=\mathscr{I}m(\boldsymbol{S}_1)=\mathscr{I}m(\hat{S}_1\,.\,e^{j\phi}\,.\,e^{j\omega t}) \tag{1.6}$$

and

$$s_2=\hat{S}_2\,.\,\sin\omega t=\mathscr{I}m(\boldsymbol{S}_2)=\mathscr{I}m(\hat{S}_2\,.\,e^{j\omega t}) \tag{1.7}$$

when the positive phase angle in the expression for s_1 indicates that s_1 is leading s_2.

Referring to equations (1.2) to (1.7) it is apparent that time dependence can be eliminated by eliminating the unit vector from these equations. The resulting equations are in **phasor** form and can be mapped on to the complex plane of the **Argand diagram.**

Phasors are represented on the Argand diagram by a complex number defining the position of their end point in the plane. For example, the signals s_1 and s_2 defined in equations (1.6) and (1.7) would be represented in **peak phasor** form ($\hat{\boldsymbol{S}}_1$ and $\hat{\boldsymbol{S}}_2$) by complex numbers as follows:

$$\hat{\boldsymbol{S}}_1 = \hat{S}_1 . (\cos \phi + \mathrm{j} . \sin \phi) \tag{1.8}$$

and

$$\hat{\boldsymbol{S}}_2 = \hat{S}_2 . (\cos 0 + \mathrm{j} . \sin 0) \tag{1.9}$$

and can be plotted on the Argand diagram as in Fig. 1.4.

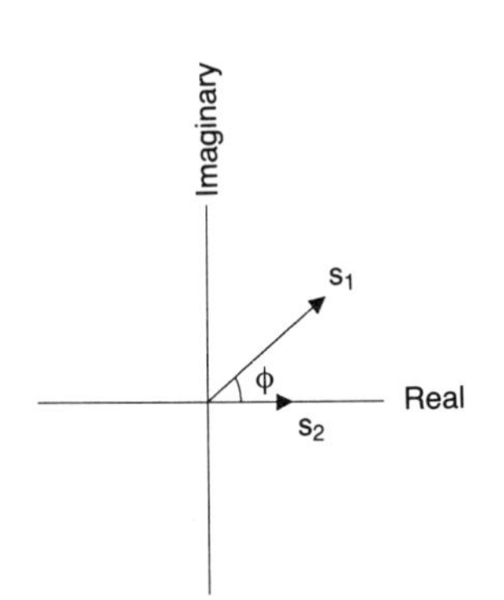

Fig. 1.4 Argand diagram for $\hat{\boldsymbol{S}}_1$ and $\boldsymbol{S}_2$ phasors.

Comment

As phasor $\hat{\boldsymbol{S}}_2$ is the reference phasor it is wholly real.

Alternatively, phasors can be expressed using the **complex polar form** as in equations (1.10) and (1.11).

$$\hat{\boldsymbol{S}}_1 = \hat{S}_1 \underline{|\phi} \tag{1.10}$$

and

$$\hat{\boldsymbol{S}}_2 = \hat{S}_2 \underline{|0^\circ} \tag{1.11}$$

Comment

As in equation (1.11) phasor $\hat{\boldsymbol{S}}_2$ is wholly real, the expression for the angle could be omitted from the equation which would then have the form

$$\hat{\boldsymbol{S}}_2 = \hat{S}_2$$

in which the zero angle is understood.

Similar expressions could have been derived for equations (1.2) to (1.5).

Working from the Argand diagram, complex algebra can then be used to calculate the amplitude and relative phase relationships of all phasors in any given system as shown in detail in Appendix A.

1.2 Complex impedances

The relationship between the **resistive** (R), **inductive** (L) and **capacitive** (C) components of impedance and the associated current and voltage phasors is important to the analysis of a.c. systems.

Consider the series RLC circuit of Fig. 1.5 in series with a current generator delivering a sinusoidal current of $i = \hat{I} \sin \omega t$. At steady state, the instantaneous a.c. voltages v_R, v_L and v_C across the R, L and C components respectively can be written as:

$$v_R = i . R = R . \hat{I} . \sin \omega t = \hat{V}_R . \sin \omega t \tag{1.12}$$

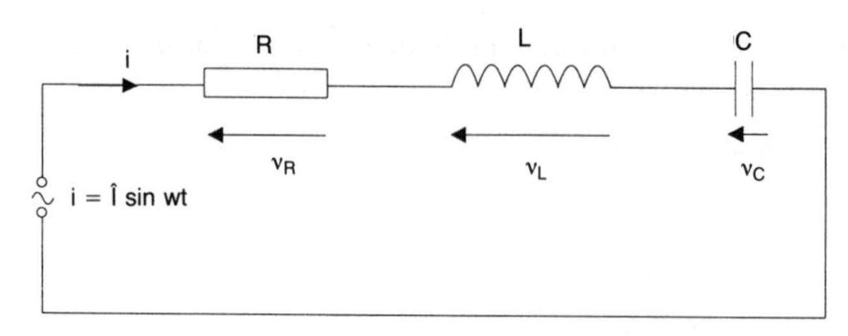

Fig. 1.5 Series *RLC* circuit.

$$v_L = L.\frac{di}{dt} = L.\omega.\hat{I}.\cos\omega t = \hat{V}_L.\cos\omega t \tag{1.13}$$

$$v_C = \frac{1}{C}\int i.dt = -\frac{\hat{I}}{\omega C}.\cos\omega t = -\hat{V}_C.\cos\omega t \tag{1.14}$$

Expressed in peak phasor form with current as reference gives:

$$\hat{\boldsymbol{V}}_R = \hat{V}_R\underline{|0°} = \hat{V}_R \tag{1.15}$$

$$\hat{\boldsymbol{V}}_L = \hat{V}_L\underline{|90°} = j.\hat{V}_L \tag{1.16}$$

and

$$\hat{\boldsymbol{V}}_C = \hat{V}_C\underline{|-90°} = -j.\hat{V}_C \tag{1.17}$$

The effective impedance ($\boldsymbol{Z}$) of the circuit is then:

$$\boldsymbol{Z} = R + j\omega L + \frac{1}{j\omega C} = R + j.\left(\omega L - \frac{1}{\omega C}\right) = Z\underline{|\phi} \tag{1.18}$$

where

$$Z = |\boldsymbol{Z}| = \left[R^2 + \left(\omega L - \frac{1}{\omega C}\right)^2\right]^{\frac{1}{2}}$$

is the magnitude of the impedance
and

$$\phi = \tan^{-1}\left[\frac{\left(\omega L - \frac{1}{\omega C}\right)}{R}\right]$$

is the phase angle of the impedance

Comment

As $\boldsymbol{Z}$ does not vary with time it is a **complex impedance** and is not a phasor.

The phasor expression for the circuit voltage then becomes:

$$\hat{V} = \hat{I}\,.\,Z = \hat{I}\,.\left[R + j\,.\left(\omega L - \frac{1}{\omega C}\right)\right] \tag{1.19}$$

The relationships of equations (1.15), (1.16), (1.17) and (1.19) can be plotted on the Argand diagram to give the phasor diagram of the series *RLC* circuit using peak phasors is then as shown in Fig. 1.6. Note that it is common to plot both the current and voltage phasors on the same diagram, in which case care must be taken that appropriate scales are used for each on both the real and imaginary axes.

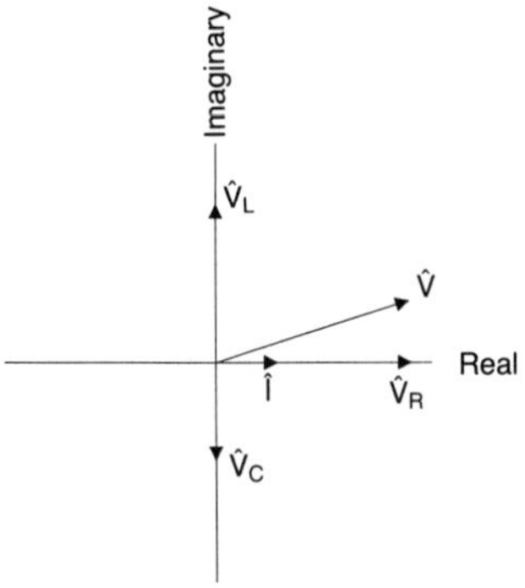

Fig. 1.6 Phasor diagram for the series *RLC* circuit.

Example 1.1 Circuit analysis using phasors
Consider the circuits of Fig. 1.7 and Fig. 1.10 in which it is required to find the value of the voltage phasor $\hat{V}_{ab}$.

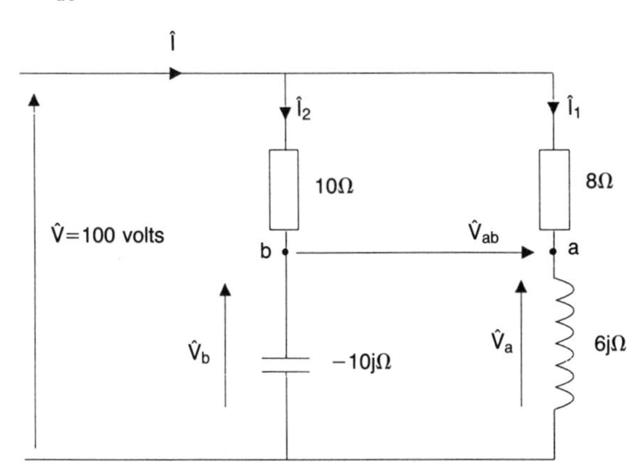

Fig. 1.7 Circuit for Example 1.1.

(i) By drawing the phasor diagrams
Choosing the phasor of the source voltage ($\hat{V}$) as reference then:

$$\hat{I}_1 = \frac{\hat{V}}{Z_1} = 10\underline{|-36.87^\circ} \text{ amperes}$$

and

$$\hat{I}_2 = \frac{\hat{V}}{Z_2} = 7.07\underline{|45^\circ} \text{ amperes}$$

since

$$Z_1 = 8 + 6\text{j} = 10\underline{|36.87^\circ} \text{ ohms}$$

and

$$Z_2 = 10 - 10\text{j} = 14.14\underline{|-45^\circ} \text{ ohms}$$

The phasor diagram for $\hat{I}_1$, $\hat{I}_2$ and the $\hat{I}$ phasors can now be constructed (Fig. 1.8).

The full phasor diagram combining both currents and voltages is then as shown in Fig. 1.9 from which the magnitude and direction of the $\hat{V}_{ab}$ phasor can be obtained.

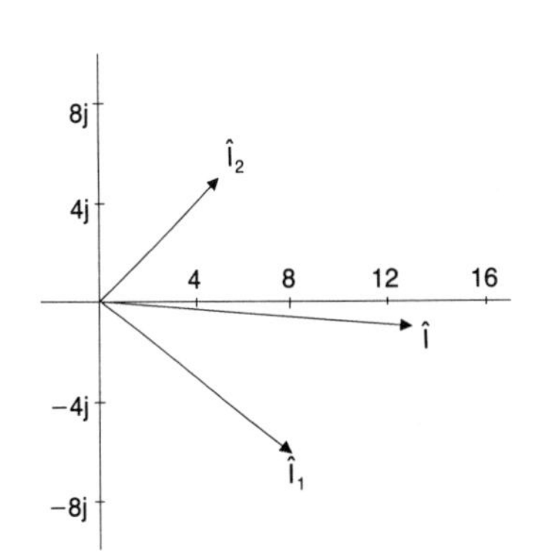

Fig. 1.8 Current phasors for Example 1.1.

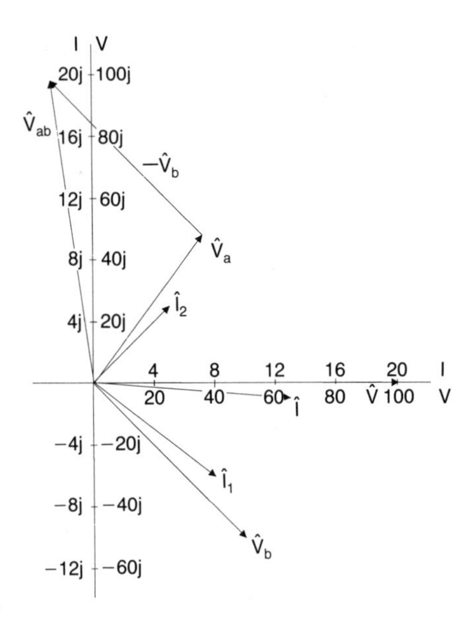

Fig. 1.9 Voltage and current phasors for Example 1.1.

Interchanging the 10 Ω and $-10j$ Ω impedances gives the circuit of Fig. 1.10 in which the values of $\hat{\boldsymbol{I}}_1$, $\hat{\boldsymbol{I}}_2$ and $\hat{\boldsymbol{I}}$ are all unchanged. Redrawing the phasor diagram enables the new value of voltage $\hat{\boldsymbol{V}}_{ab}$ to be found as in Fig. 1.11.

(ii) By calculation

From the configuration of the circuit:

$$\hat{\boldsymbol{I}} = \hat{\boldsymbol{I}}_1 + \hat{\boldsymbol{I}}_2 = 13 - j = 13.04\underline{|-4.4^\circ} \text{ A}$$

For the arrangement of Fig. 1.7

$$\hat{\boldsymbol{V}}_a = 6j \, . \, \hat{\boldsymbol{I}}_1 = 60\underline{|53.13^\circ} = 36 + 48j \text{ V}$$

and

$$\hat{\boldsymbol{V}}_b = -10j \, . \, \hat{\boldsymbol{I}}_2 = 70.71\underline{|-45^\circ} = 50 - 50j \text{ V}$$

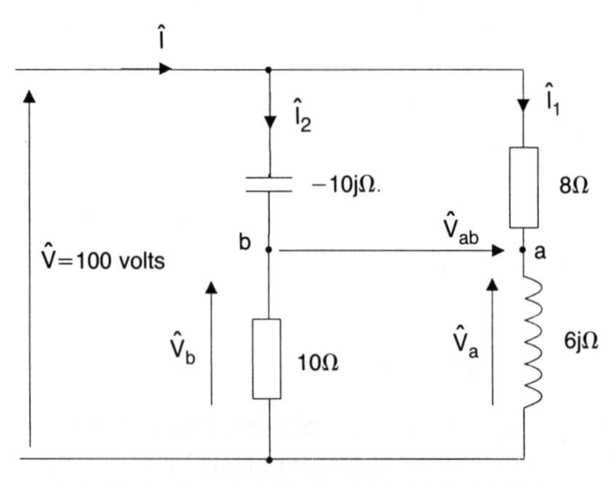

Fig. 1.10 Revised circuit form for Example 1.1.

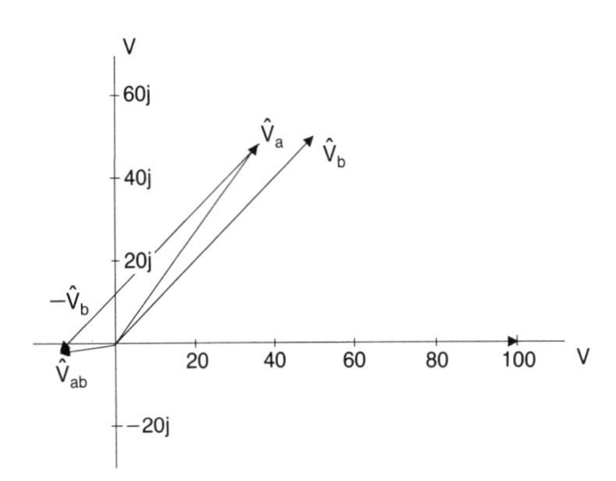

Fig. 1.11 Voltage phasors for revised circuit form of Fig. 1.10.

when

$$\hat{V}_{ab}=\hat{V}_a-\hat{V}_b=-14+98j=98.99\underline{|98.13^\circ}\ \text{V}$$

For the arrangement of Fig. 1.10

$$\hat{V}_a=6j\,.\,\hat{I}_1=60\underline{|53.13^\circ}=36+48j\ \text{V}$$

and

$$\hat{V}_b=10\,.\,\hat{I}_2=70.71\underline{|45^\circ}=50+50j\ \text{V}$$

when

$$\hat{V}_{ab}=\hat{V}_a-\hat{V}_b=-14-2j=14.14\underline{|-171.87^\circ}\ \text{V}$$

1.3 Power in a.c. circuits

If a circuit consisting solely of a resistor R has a current $i=\hat{I}\,.\,\sin\omega t$ flowing through it then the voltage across R will be:

$$v=\hat{I}\,.\,R\,.\,\sin\omega t=\hat{V}\,.\,\sin\omega t \tag{1.20}$$

Comment

Energy is only dissipated in the resistive elements of a circuit.

The instantaneous power dissipated in the resistor will then be:

$$P_{\text{inst}}=v\,.\,i=\hat{V}\,.\,\hat{I}\,.\,\sin^2\omega t$$

$$=\frac{\hat{V}\,.\,\hat{I}}{2}(1-\cos 2\omega t)=V\,.\,I(1-\cos 2\omega t) \tag{1.21}$$

In which V and I represent the **root mean square** or r.m.s. values of the voltage and current such that:

$$V = \frac{\hat{V}}{\sqrt{2}} \tag{1.22}$$

and

$$I = \frac{\hat{I}}{\sqrt{2}} \tag{1.23}$$

Over one cycle this gives an average power of:

$$P = V \,.\, I = I^2 \,.\, R \tag{1.24}$$

Comment

In general, the parameters of a.c. power systems are described in terms of the r.m.s. values of their currents and voltages and not by peak values. The effect on the previous discussion is simply to substitute the r.m.s. phasor for the peak phasor considered thus far. Referring to equation (1.19), the same expression using r.m.s. phasors would be:

$$\boldsymbol{V} = \boldsymbol{I} \,.\, \boldsymbol{Z} = \boldsymbol{I} \,.\left[R + j \,.\left(\omega L - \frac{1}{\omega C}\right)\right]$$

in which the only difference is the use of the r.m.s. phasors $\boldsymbol{V}$ and $\boldsymbol{I}$ instead of the peak phasors $\hat{\boldsymbol{V}}$ and $\hat{\boldsymbol{I}}$.

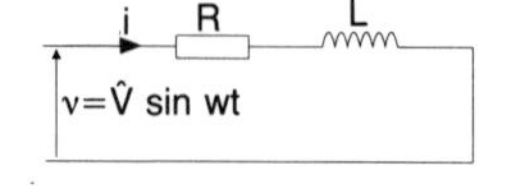

Fig. 1.12 Series RL circuit.

Consider now the series RL circuit of Fig. 1.12, the r.m.s. phasor diagram for which is shown in Fig. 1.13.

Comment

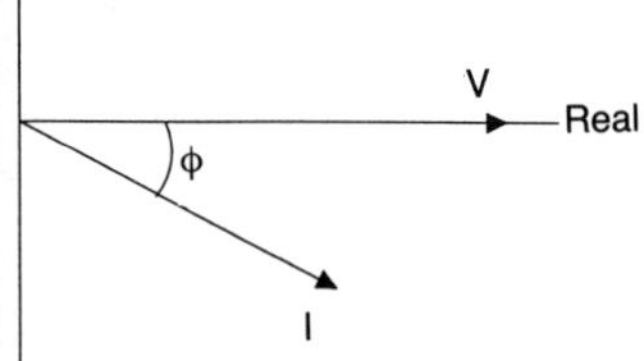

Fig. 1.13 Phasor diagram for series RL circuit.

The phase angle ϕ in Fig. 1.13 is obtained from the values of L and R and the supply frequency as:

$$\phi = \tan^{-1}\left(\frac{\omega L}{R}\right)$$

Assuming a voltage of $v = \hat{V} \sin \omega t$ and a current of $i = \hat{I} \sin(\omega t - \phi)$ in the circuit, the instantaneous power delivered is given by

$$P_{\text{inst}} = v \,.\, i = \hat{V} \,.\, \hat{I} \,.\, \sin \omega t \,.\, \sin(\omega t - \phi) = \frac{\hat{V} \,.\, \hat{I}}{2}\,[\cos \phi - \cos(2\omega t - \phi)] \tag{1.25}$$

This gives an average power over one cycle of:

$$P = V \,.\, I \,.\cos \phi \tag{1.26}$$

The VI product in equation (1.26) is referred to as the **apparent power** and is given the symbol S. Though having the same dimensions as power, the apparent power is not usually expressed in watts but is given the units of **volt-amperes** or VA. The cos ϕ term in equation (1.26) is referred to as the **power factor** and is a measure of the reactive component of the load

---

Comment

In the case where the circuit is purely resistive $\cos\phi = 1$.

With reference to Fig. 1.13, the $V\cos\phi$ term is seen to correspond to the voltage across the resistor. Hence:

$$P = V.I.\cos\phi = I^2R \qquad (1.27)$$

Similarly, the voltage across the inductor is $V\sin\phi$. Multiplying this expression by the current gives the **reactive power** in the system, defined by the symbol Q such that:

$$Q = V.I.\sin\phi \qquad (1.28)$$

As with apparent power, reactive power has the dimensions of watts but is given the units of **volt-amperes reactive** or VAR. Inductors are VAR lagging and are considered to absorb VAR while capacitors are VAR leading and are considered as a source of VAR.

Comment

Power system components – motors, transformers, capacitors, etc. – have a **current rating** which defines the maximum continuous current that they can draw without suffering damage. Conventionally, this current rating is expressed in terms of a **volt-ampere rating** at a specified value of supply voltage.

P, Q and S can be related using the power triangle developed from the voltage and current phasor diagram of Fig. 1.14 on which are also drawn the **inphase** or **direct** ($I_d = I\cos\phi$) and **quadrature** ($I_q = I\sin\phi$) components of the current phasor $\boldsymbol{I}$.

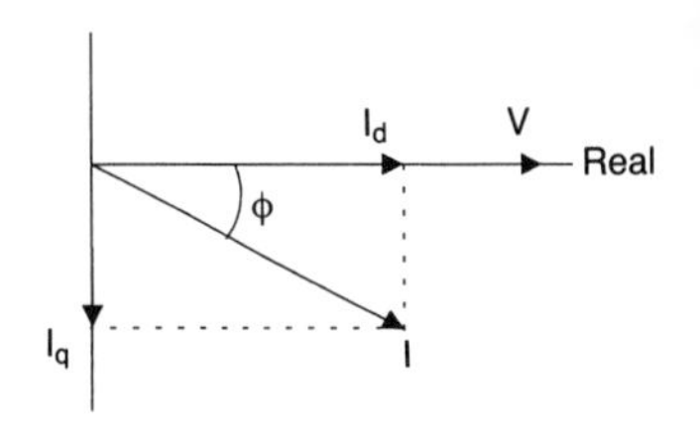

Fig. 1.14 Phasor diagram showing direct (I_d) and quadrature (I_q) components of current.

Comment

If the power, and hence $I\cos\phi$, remained the same while the angle ϕ was reduced then I would also reduce as $\cos\phi$ increases with decreasing ϕ.

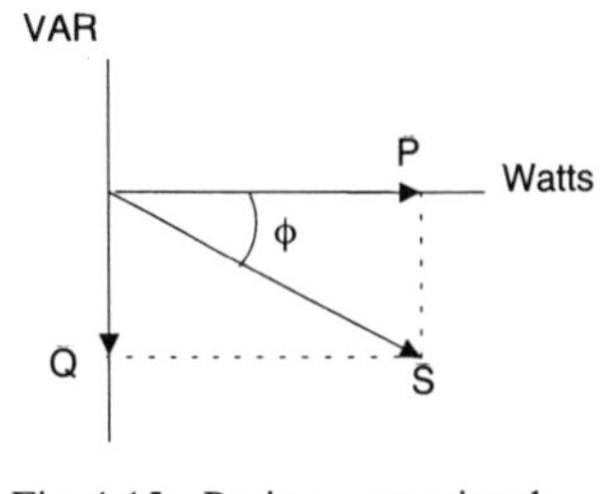

Fig. 1.15 Basic power triangle.

Multiplying $\boldsymbol{I}$, $I\cos\phi$ and $I\sin\phi$ by V yields the result of Fig. 1.15 which is one form of the **power triangle** in which $S = VI$, $P = VI.\cos\phi$ and $Q = VI.\sin\phi$.

Comment

P, Q and S in Fig. 1.15 are not phasor quantities as they do not represent sinusoidal, time varying parameters. However, like phasors, they can be manipulated in complex form using complex arithmetic.

For the form of the power triangle represented by Fig. 1.15 with $\boldsymbol{V}$ as reference:

$$S=\boldsymbol{V}.\boldsymbol{I}=V\underline{|0^\circ}.I\underline{|-\phi}=V.I(\cos\phi-\mathrm{j}.\sin\phi)=P-\mathrm{j}Q \tag{1.29}$$

This, however, has the unfortunate consequence of representing the reactive power associated with the inductor as negative whereas the accepted convention stated earlier associates an inductor with positive VAR!

A further problem arises if for any reason the voltage phasor cannot be expressed as $V\underline{|0^\circ}$. Suppose that a voltage phasor $\boldsymbol{V}=V\underline{|\theta}$ was to be used, the corresponding current phasor for a current lagging the voltage by an angle ϕ would then be $\boldsymbol{I}=I\underline{|\theta-\phi}$.

Forming S as $\boldsymbol{VI}$ gives in this instance:

$$S=\boldsymbol{V}.\boldsymbol{I}=V\underline{|\theta}.I\underline{|(\theta-\phi)}=V.I\underline{|(2\theta-\phi)} \tag{1.30}$$

referring to equations (1.26), (1.28) and (1.29), for a phase angle of ϕ between the current and voltage phasors the expression for S should have the form:

$$S=V.I.\cos\phi+\mathrm{j}.V.I.\sin\phi \tag{1.31}$$

whereas equation (1.30) gives:

$$S=V.I.\cos(2\theta-\phi)+\mathrm{j}.V.I.\sin(2\theta-\phi) \tag{1.32}$$

which is obviously wrong!

To correct this the $\boldsymbol{I}$ phasor is replaced by its conjugate $\boldsymbol{I}^*$ such that:

$$\boldsymbol{I}^*=I\underline{|-(\theta-\phi)} \tag{1.33}$$

in which case:

$$S=\boldsymbol{V}.\boldsymbol{I}^*=V\underline{|t}.I\underline{|-(\theta-\phi)}=V.I\underline{|\phi}=P+\mathrm{j}Q \tag{1.34}$$

This then gives a power triangle of the form of Fig. 1.16 which is now in the correct sense for both inductive and capacitive VAR but is a 'mirror image' of that derived directly from the phasor diagram.

This, however, is the form commonly used!

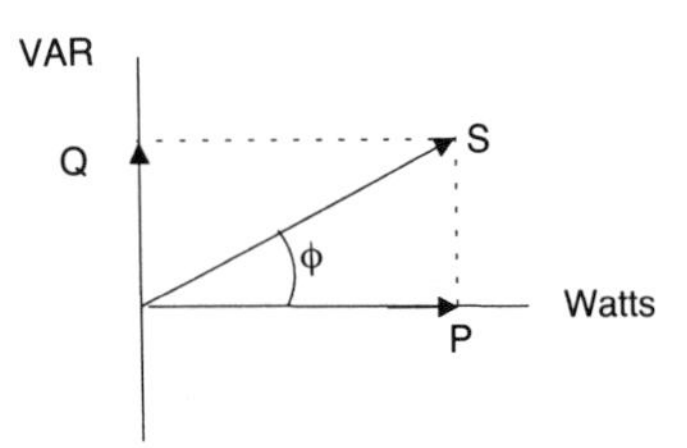

Fig. 1.16 Corrected power triangle.

1.3.1 *Power factor correction*

For a given energy transfer in a typical power system with a fixed supply voltage the current required depends upon the power factor of the load. The supplier of the energy has to send this current through the transmission network in which there will be an inevitable system power loss due to resistance (I^2R loss). It is therefore an advantage to the supplier for the system currents to be as low as possible.

A consumer wants to pay only for energy supplied but the supplier will also charge large consumers for VAR in order to compensate for the increased system power losses

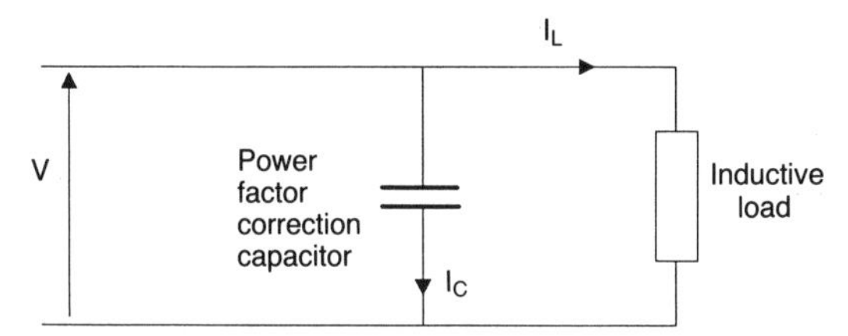

Fig. 1.17 Power factor correction capacitors.

associated with low power factor loads. It is therefore advantageous for the consumer to ensure that the load is kept as close to unity power factor as possible. As most industrial loads are inductive, this can be achieved by introducing **power factor correction capacitors** in parallel with the main load as in Fig. 1.17 to supply the necessary VAR.

The amount of capacitance required to bring the power factor to unity can be determined by calculation – resulting in the effective impedance as seen by the power system becoming entirely resistive – or by superposition of the power triangles for the load and the capacitor, respectively.

For the inductive load of Fig. 1.18 the power triangle, using the $\boldsymbol{V}.\boldsymbol{I}^*$ form, is as shown in Fig. 1.19. The power triangle for the power factor correction capacitor is as shown by Fig. 1.20. If the value of Q for the inductive circuit is the same as that for the power factor correction capacitors the combined power triangle will reduce to that of Fig. 1.21 which contains the real power term only. The effective power factor as seen by the power system is then unity. The phasor diagram for the system is shown in Fig. 1.22. If full correction is applied then $\boldsymbol{V}$ and $\boldsymbol{I}_L$ will be in phase.

In practice, as the characteristics of the load change over the course of a day a corresponding change in the value of power factor correction capacitance will also be required if the effective load power factor is to be maintained at unity.

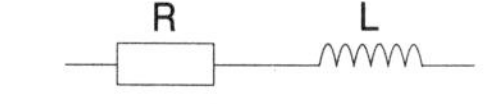

Fig. 1.18 Inductive load.

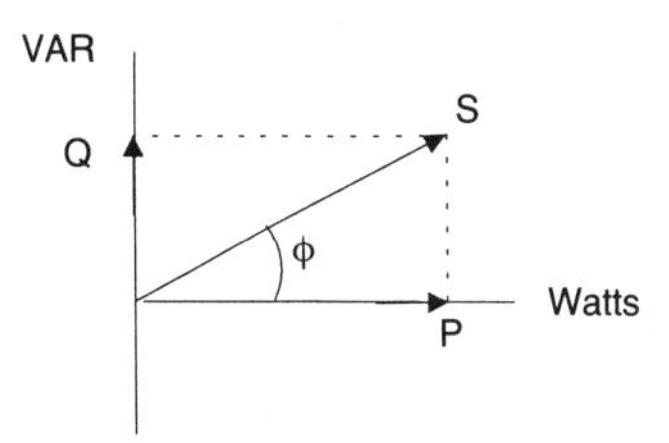

Fig. 1.19 Power triangle for an inductive load.

Example 1.2 Power factor correction

An inductive load of 100.18 $\underline{|26.05^\circ}$ ohms is connected to a 63 500 volt supply. What value of capacitor, expressed both as an appropriate VAR value and in ohms, is required to provide an effective power factor of unity at the terminals of the supply.

Current in load:

$$I = \frac{63\,500}{100.18}\underline{|-26.05^\circ} = 633.86\underline{|-26.05^\circ} \text{ amperes}$$

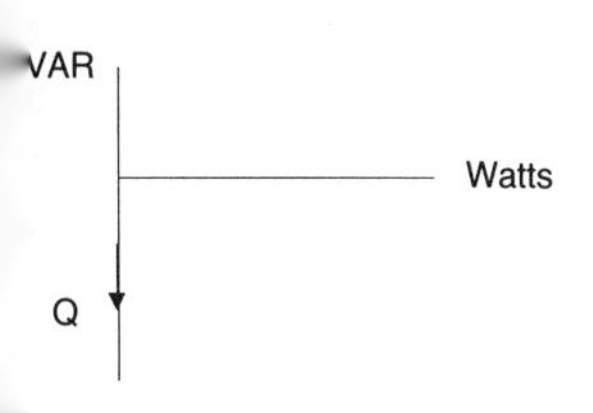

Fig. 1.20 Power triangle for capacitor.

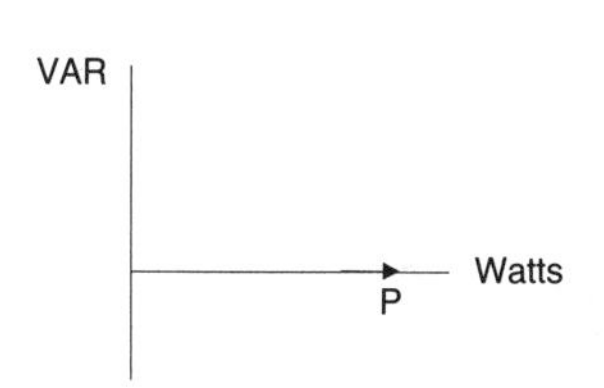

Fig. 1.21 Power triangle formed from the combination of Figs 1.19 and 1.20.

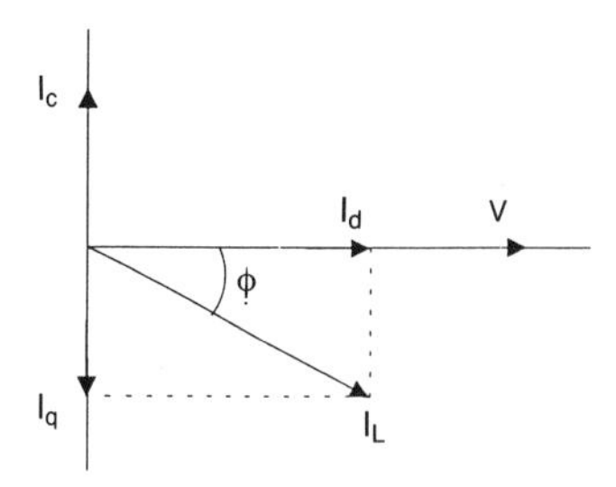

Fig. 1.22 Phasor diagram for the circuit of Fig. 1.17.

The volt-ampere rating of the load is then:

$$S = V \,.\, I^* = (36.16 + 17.68\mathrm{j}) \times 10^6$$

and

$$|S| = 40.25 \times 10^6 \text{ VA}$$

This gives the power triangle of Fig. 1.23. Hence require 17.68 MVAR of power factor correction capacitor which is equivalent to a capacitive reactance (X_c) of:

$$X_c = \frac{63\,500^2}{17.68 \times 10^6} = 228.1 \text{ ohms}$$

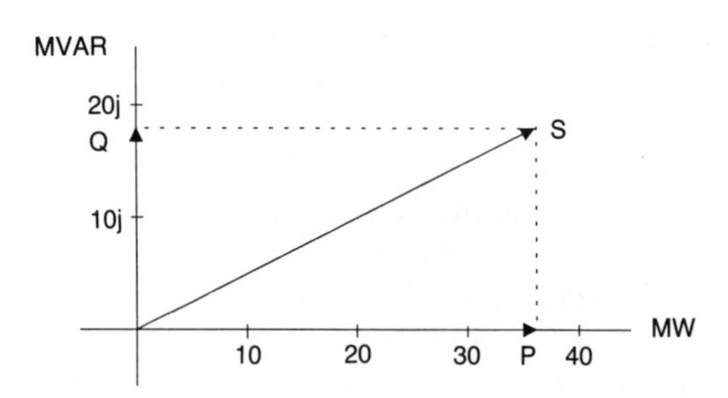

Fig. 1.23 Power triangle for Example 1.2.

1.4 Three-phase systems

By virtue of their construction, requiring a single conductor for each phase with a common or earth return, **three-phase** systems tend to be significantly cheaper than single-phase systems for the transmission of electrical power. In addition, a three-phase system facilitates torque production in an a.c. machine while also making better use of the space available for windings in such machines.

The balanced three-phase source of Fig. 1.24 has instantaneous phase voltages referred to the reference **neutral** of:

$$\begin{vmatrix} v_1 \\ v_2 \\ v_3 \end{vmatrix} = \hat{V}_p \begin{vmatrix} \sin \omega t \\ \sin\left(\omega t - \dfrac{2\pi}{3}\right) \\ \sin\left(\omega t - \dfrac{4\pi}{3}\right) \end{vmatrix} \tag{1.35}$$

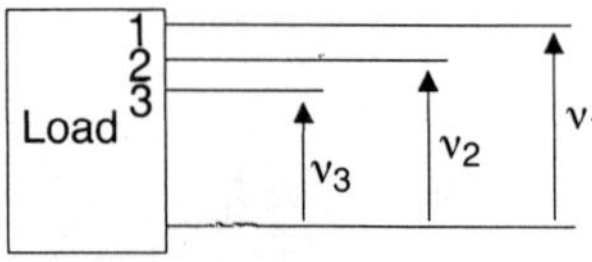

Fig. 1.24 Balanced three-phase source.

and

$$v_1 + v_2 + v_3 = 0 \tag{1.36}$$

or, in terms of the r.m.s. phasors:

$$\begin{vmatrix} \boldsymbol{V}_1 \\ \boldsymbol{V}_2 \\ \boldsymbol{V}_3 \end{vmatrix} = V_p \begin{vmatrix} \underline{|0} \\ \underline{\left| -\frac{2\pi}{3} \right.} \\ \underline{\left| -\frac{4\pi}{3} \right.} \end{vmatrix} \tag{1.37}$$

and

$$\boldsymbol{V}_1 + \boldsymbol{V}_2 + \boldsymbol{V}_3 = 0 \tag{1.38}$$

The relationship between these phasors is as shown in Fig. 1.25.

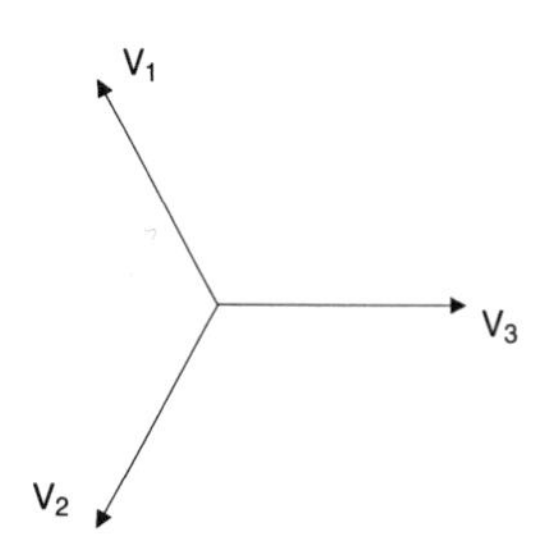

Fig. 1.25 Phasor diagram for a balanced three-phase set of voltages.

If this source is connected to a **balanced load** the corresponding currents will also form a balanced set such that:

$$\begin{vmatrix} i_1 \\ i_2 \\ i_3 \end{vmatrix} = \hat{I}_p \begin{vmatrix} \sin(\omega t - \phi) \\ \sin\left(\omega t - \frac{2\pi}{3} - \phi\right) \\ \sin\left(\omega t - \frac{4\pi}{3} - \phi\right) \end{vmatrix} \tag{1.39}$$

and

$$\begin{vmatrix} \boldsymbol{I}_1 \\ \boldsymbol{I}_2 \\ \boldsymbol{I}_3 \end{vmatrix} = I_p \begin{vmatrix} \underline{|-\phi} \\ \underline{\left| -\left(\frac{2\pi}{3} + \phi\right) \right.} \\ \underline{\left| -\left(\frac{4\pi}{3} + \phi\right) \right.} \end{vmatrix} \tag{1.40}$$

Comment

In a balanced three-phase load each phase is presented with an identical impedance.

In general, a three-phase system is not defined by its phase voltage but by the voltage between pairs of lines (the **line voltages**) such that, referring to Fig. 1.26(a).

$$\begin{vmatrix} v_{12} \\ v_{23} \\ v_{31} \end{vmatrix} = \begin{vmatrix} v_1 - v_2 \\ v_2 - v_3 \\ v_3 - v_1 \end{vmatrix} \tag{1.41}$$

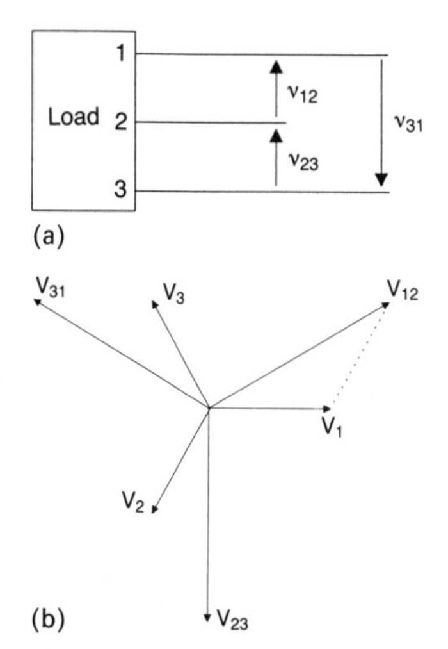

Fig. 1.26 Line voltage relationships: (a) line voltages; and (b) line voltage phasors.

The relationship between the line voltage phasors is then as illustrated by Fig. 1.26(b) from which it is seen that the line voltages also form a balanced set such that, with v_1, the voltage in phase l, as reference:

$$\begin{vmatrix} v_{12} \\ v_{23} \\ v_{31} \end{vmatrix} = \hat{V}_p \begin{vmatrix} \sin \omega t - \sin\left(\omega t - \frac{2\pi}{3}\right) \\ \sin\left(\omega t - \frac{2\pi}{3}\right) - \sin\left(\omega t - \frac{4\pi}{3}\right) \\ \sin\left(\omega t - \frac{4\pi}{3}\right) - \sin \omega t \end{vmatrix} = \hat{V}_l \begin{vmatrix} \sin\left(\omega t + \frac{\pi}{6}\right) \\ \sin\left(\omega t - \frac{\pi}{2}\right) \\ \sin\left(\omega t - \frac{7\pi}{6}\right) \end{vmatrix} \tag{1.42}$$

where

$$\hat{V}_l = \sqrt{3}\, \hat{V}_p \tag{1.43}$$

In terms of the r.m.s. phasors, with phase l as reference:

$$\begin{vmatrix} \boldsymbol{V}_{12} \\ \boldsymbol{V}_{23} \\ \boldsymbol{V}_{31} \end{vmatrix} = V_l \begin{vmatrix} \angle \frac{\pi}{6} \\ \angle -\frac{\pi}{2} \\ \angle -\frac{7\pi}{6} \end{vmatrix} \tag{1.44}$$

and

$$V_l = \sqrt{3} \,.\, V_p \tag{1.45}$$

1.4.1 *Power in a balanced three-phase system*

The instantaneous three-phase power in the balanced sytem is obtained as the sum of the instantaneous powers in each of the three phases such that:

$$P_{\text{inst}} = v_1 \,.\, i_1 + v_2 \,.\, i_2 + v_3 \,.\, i_3 \tag{1.46}$$

Using equations (1.33) and (1.37) gives an expression for the average power over one cycle of:

$$P = 3V_p \,.\, I_p \,.\cos \phi \text{ watts} \tag{1.47}$$

in which I_p and V_p are the r.m.s. values of the phase current and phase voltage respectively, while cos ϕ is the phase angle of the load impedance and is therefore the angle between the corresponding pairs of phase current and phase voltage.

Alternatively:

$$S = \mathscr{R}(\boldsymbol{V}_1 \,.\, \boldsymbol{I}_1^* + \boldsymbol{V}_2 \,.\, \boldsymbol{I}_2^* + \boldsymbol{V}_3 \,.\, \boldsymbol{I}_3^*) = P = 3V_p \,.\, I_p \,.\cos \phi \tag{1.48}$$

Comment

$\mathscr{R}(\boldsymbol{V}_1\boldsymbol{I}_1^* + \boldsymbol{V}_2\boldsymbol{I}_2^* + \boldsymbol{V}_3\boldsymbol{I}_3^*)$ is the real part of the sum of the complex quantities $\boldsymbol{V}_1\boldsymbol{I}_1^*$, etc.

1.4.2 *Balanced three-phase loads*

In a balanced three-phase load the individual impedances making up the load each have the same magnitude and phase angle.

(a) *Balanced four-wire star connected load*

In the balanced four-wire star of Fig. 1.27 the impedances in each leg of the load are identical while the appropriate phase voltage appears across the associated load. The line current ($\boldsymbol{I}_l$) is the same as the phase current ($\boldsymbol{I}_p$) while the line voltage appears across pairs of impedances.

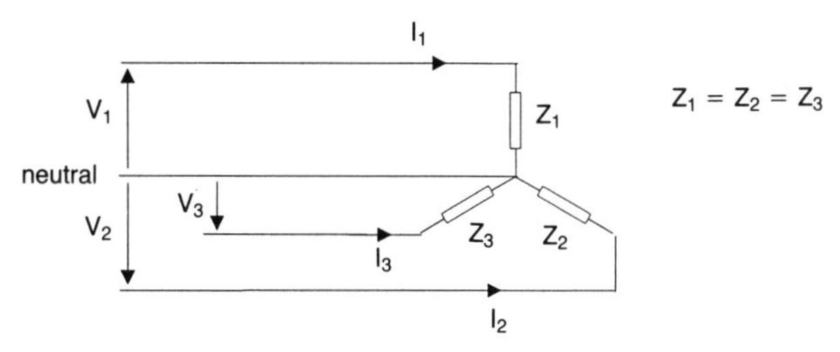

Fig. 1.27 Balanced three-phase, four-wire star connected load.

The relationship between the individual phase currents and voltages in the balanced four-wire star connected load is:

$$\begin{vmatrix} \boldsymbol{V}_1 \\ \boldsymbol{V}_2 \\ \boldsymbol{V}_3 \end{vmatrix} = \boldsymbol{Z}_Y \begin{vmatrix} \boldsymbol{I}_1 \\ \boldsymbol{I}_2 \\ \boldsymbol{I}_3 \end{vmatrix} \tag{1.49}$$

Since these currents form a balanced set the current in the fourth or neutral conductor is:

$$\boldsymbol{I}_1 + \boldsymbol{I}_2 + \boldsymbol{I}_3 = 0 \tag{1.50}$$

The total power in the load is:

$$P = 3V_p I_p \cos\phi \tag{1.51}$$

also, since $V_l = \sqrt{3}\,V_p$ and $I_l = I_p$

$$P = \sqrt{3}\,V_l . I_l . \cos\phi \tag{1.52}$$

Comment

The angle ϕ in equation (1.52) is the phase angle of the phase impedance of the load $(Z_Y\underline{|\phi})$ and is not the angle between the line voltage (V_l) and the line current (I_l).

(b) *Balanced three-wire star connected load*

Referring to equation (1.50) it is seen that no current flows in the fourth or neutral conductor of the balanced four-wire star connected load and this conductor can therefore be removed to give the balanced three-wire star without affecting the relationships of equations (1.49) and (1.50) or the power as expressed by equations (1.51) and (1.52). These equations therefore apply to the balanced three-wire star connected load of Fig. 1.28.

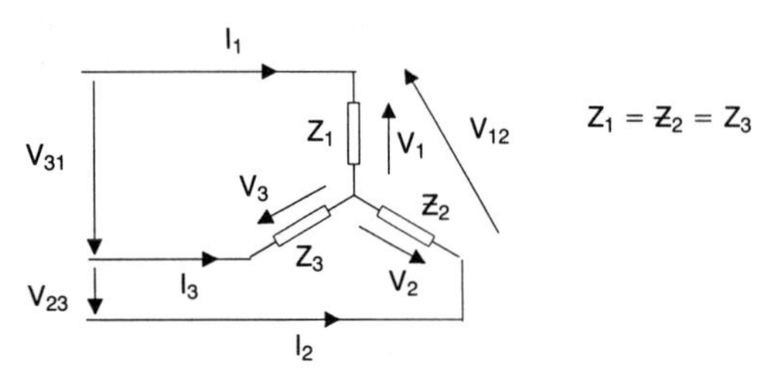

Fig. 1.28 Balanced three-phase, four-wire star connected load.

(c) *Balanced delta connected load*

For the balanced delta load of Fig. 1.29 the voltage that appears across each of the load impedances is the appropriate line voltage while the phase currents are calculated from the relationships:

$$\begin{vmatrix} V_{12} \\ V_{23} \\ V_{31} \end{vmatrix} = Z_D \begin{vmatrix} I_{12} \\ I_{23} \\ I_{31} \end{vmatrix} \quad (1.53)$$

and

$$\begin{vmatrix} I_1 \\ I_2 \\ I_3 \end{vmatrix} = \begin{vmatrix} I_{12}-I_{31} \\ I_{23}-I_{12} \\ I_{31}-I_{23} \end{vmatrix} \quad (1.54)$$

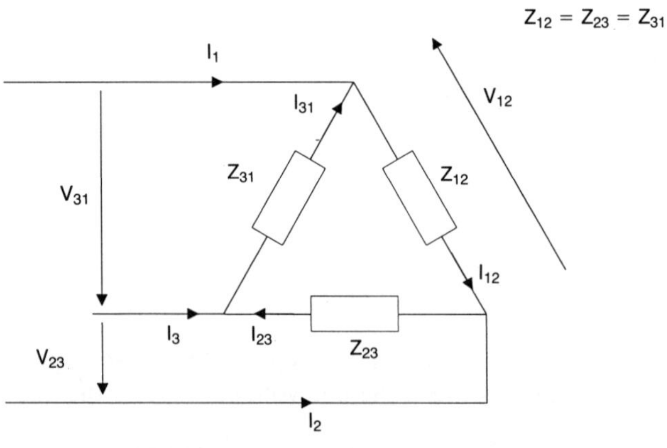

Fig. 1.29 Balanced delta connected load.

Referring to Fig. 1.29, the relationship between the line and phase current is then:

$$I_l = \sqrt{3}\, I_p \tag{1.55}$$

The three-phase power is:

$$P = 3V_p \,.\, I_p \,.\, \cos\phi \tag{1.56}$$

or, in terms of line current and voltage:

$$P = \sqrt{3}\, V_l \,.\, I_l \,.\, \cos\phi \tag{1.57}$$

Comment

The angle ϕ in equation (1.57) is the phase angle of the load impedance ($Z_D\underline{|\phi}$) and is not the angle between the line voltage (V_l) and the line current (I_l).

Example 1.3 The balanced delta connected load
For a balanced delta connected load of impedance $Z_D = (10 + 4j)$ find the phase and line currents and the power taken by the load.

The line voltage phasor equations are:

$$\begin{vmatrix} V_{12} \\ V_{23} \\ V_{31} \end{vmatrix} = 415 \begin{vmatrix} \underline{|0^\circ} \\ \underline{|-120^\circ} \\ \underline{|-240^\circ} \end{vmatrix} = \begin{vmatrix} 415 + 0j \\ -207.5 - 359.4j \\ -207.5 + 359.4j \end{vmatrix} = (10 + 4j) \begin{vmatrix} I_{12} \\ I_{23} \\ I_{31} \end{vmatrix} = 10.77\underline{|21.8^\circ} \begin{vmatrix} I_{12} \\ I_{23} \\ I_{31} \end{vmatrix}$$

Solving for I_{12} gives:

$$I_{12} = \frac{415}{10.77}\underline{|-21.8^\circ} = 38.53\underline{|-21.8^\circ} = 35.78 - 14.31j \text{ A}$$

Hence

$$I_{23} = 38.53\underline{|-141.8^\circ}$$

and

$$I_{31} = 38.53\underline{|-221.8^\circ} \text{ A}$$

and

$$I_1 = I_{12} - I_{31} = \sqrt{3} \,.\, I_{12} \,.\, \underline{|-51.8^\circ} = 66.74\underline{|-51.8^\circ} = 41.27 - 55.45j \text{ A}$$

when

$$I_2 = 66.74\underline{|-171.8^\circ}$$

and

$$I_3 = 66.74\underline{|-291.8^\circ} \text{ A}$$

Then

$$\text{Power} = \sqrt{3} \times 415 \times 66.74 \times \cos(21.8^\circ) = 44\,540 \text{ watts}$$

or

$$\text{Power} = 3 \,.\, I_{12}^2 \,.\, R = 3 \times 38.53^2 \times 10 = 44\ 540 \text{ watts}$$

1.4.3 *Single line diagrams*

Any balanced three-phase system may be analysed by reference to a single-phase only. This representation of the three-phase system is referred to as the **representative** or **equivalent single line diagram** form. The solutions for the other two phases are equal in magnitude but shifted in phase by 120° and −120° degrees, respectively, while the three-phase power is obtained by multiplying the power in the single phase by three.

The single line diagram assumes a balanced star connected load and any balanced delta connected load must be converted into the equivalent balanced star. For the star connected load:

$$Z_Y = \frac{V_p}{I_p} = \frac{V_l}{\sqrt{3} I_l} \tag{1.58}$$

when

$$\frac{V_l}{I_l} = \sqrt{3}\, Z_Y \tag{1.59}$$

For the delta connected load:

$$Z_D = \frac{V_l}{I_p} = \frac{\sqrt{3}\, V_l}{I_l} \tag{1.60}$$

when

$$\frac{V_l}{I_l} = \frac{Z_D}{\sqrt{3}} \tag{1.61}$$

Hence, for both loads to appear as equivalent:

$$Z_Y = \frac{Z_D}{3} \tag{1.62}$$

1.4.4 *Calculation of load impedance from rating*

Power system loads are often expressed by reference to their rating in either watts or volt-amperes at a specified voltage together with their power factor. The resistive and reactive components of the load may then be calculated.

(a) *Single-phase loads*

For a single-phase load rated at P_L watts at V volts r.m.s. with a power factor of $\cos \phi$ lagging (an inductive load):

$$P_L = V \,.\, I \,.\, \cos \phi \tag{1.63}$$

when

$$\mathbf{Z}_L = \frac{\mathbf{V}}{\mathbf{I}} = \frac{V^2 \cos \phi}{P_L} (\cos \phi + \mathrm{j} \,.\, \sin \phi) \tag{1.64}$$

For a similarly rated single-phase load with a leading power factor of cos θ (a capacitive load):

$$Z_L = \frac{V}{I} = \frac{V^2 \cos \theta}{P_L} (\cos \theta - j . \sin \theta) \tag{1.65}$$

For a single phase load rated at S_L V A at V volts r.m.s. with a power factor of cos ϕ lagging (an inductive load):

$$S_L = V . I \tag{1.66}$$

when

$$Z_L = \frac{V}{I} = \frac{V^2}{S_L} (\cos \phi + j . \sin \phi) \tag{1.67}$$

For a similarly rated load with a leading power factor of cos θ (a capacitive load):

$$Z_L = \frac{V}{I} = \frac{V^2}{S_L} (\cos \theta - j . \sin \theta) \tag{1.68}$$

(b) *Balanced three-phase loads*

For a three-phase load with a three-phase rating of S_{LL} V A at a line voltage of V_l r.m.s. and a power factor of cos θ, the phase impedance for the equivalent star connected load is then obtained from equation (1.65), with $S_{LL} = 3S_L$ and $V_l = \sqrt{3}\ V_p$ as:

$$Z_L = \frac{V_l^2}{S_{LL}} (\cos \phi + j . \sin \phi) \tag{1.69}$$

for an inductive load and

$$Z_L = \frac{V_l^2}{S_{LL}} (\cos \phi - j . \sin \phi) \tag{1.70}$$

for a capacitive load.

Example 1.4 Calculation of load impedance
A three-phase load operating from an 11 000 volt (line) supply is rated at 1.2 MVA at a power factor of 0.911 lagging. Find the phase impedance of the equivalent star connected load.

Using equation (1.70):

$$Z_L = \frac{11\,000^2}{1.2 \times 10^6} . (0.911 + 0.4124j) = 91.86 + 41.58j = 100.8\underline{|24.36^\circ}\ \Omega$$

1.4.5 *Per-unit loads*

In many instances it is convenient to express system or device parameters not as absolute values but as per-unit values obtained by dividing the actual value by a base value. To enable all parameters to be expressed in per-unit terms two base values are required, usually a three-phase volt-ampere value (VA_{base}) and a line voltage value (V_{base}). The relationships between the system parameters, their per-unit value and the base values for a three-phase system are, assuming a single line representation of the system:

$$VA_{\text{base}} = \sqrt{3}\, V_{\text{base}} \cdot I_{\text{base}} \tag{1.71}$$

Therefore

$$I_{\text{base}} = \frac{VA_{\text{base}}}{\sqrt{3}\, V_{\text{base}}} \tag{1.72}$$

and

$$Z_{\text{base}} = \frac{V_{\text{base}}}{\sqrt{3}\, I_{\text{base}}} = \frac{V_{\text{base}}^2}{VA_{\text{base}}} \tag{1.73}$$

when

$$V_{\text{pu,line}} = \frac{V_{\text{line}}}{V_{\text{base}}} = V_{\text{pu,phase}} = V_{\text{pu}} \tag{1.74}$$

$$I_{\text{pu}} = \frac{I}{I_{\text{base}}} = \frac{\sqrt{3}\, V_{\text{base}} \cdot I}{VA_{\text{base}}} \tag{1.75}$$

and

$$Z_{\text{pu}} = \frac{Z}{Z_{\text{base}}} = \frac{Z \cdot VA_{\text{base}}}{V_{\text{base}}^2} \tag{1.76}$$

(a) *Change of base*

System components such as transformers and machines may well have a 'nameplat value' of per-unit impedance expressed in terms of their individual voltage and volt ampere ratings. This then needs to be converted to a per-unit value expressed in term of the system base voltage and base volt-ampere values as follows:

$$Z_{\text{pu};1} = \frac{Z}{Z_{\text{base};1}} = \frac{Z \cdot VA_{\text{base};1}}{V_{\text{base};1}^2} \tag{1.77}$$

and

$$Z_{\text{pu};2} = \frac{Z}{Z_{\text{base};2}} = \frac{Z \cdot VA_{\text{base};2}}{V_{\text{base};2}^2} \tag{1.78}$$

when

$$Z_{\text{pu};2} = Z_{\text{pu};1} \cdot \left(\frac{VA_{\text{base};2}}{VA_{\text{base};1}}\right) \cdot \left(\frac{V_{\text{base};1}}{V_{\text{base};2}}\right)^2 \tag{1.79}$$

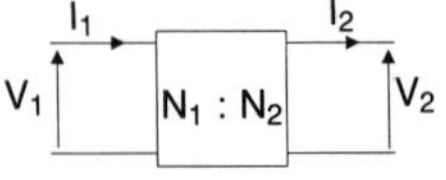

Fig. 1.30 Transformer block diagram.

(b) *The transformer in per unit form*

Figure 1.30 shows the block diagram of a transformer. For a turns ratio of N_1/N_2 th relationship between the primary and secondary voltages and currents are:

$$\frac{V_1}{V_2} = \frac{N_1}{N_2} \tag{1.80}$$

and

$$\frac{I_1}{I_2} = \frac{N_2}{N_1} \tag{1.81}$$

The base voltage may be transferred across the transformer using the relationship of equation (1.80) when, assuming that $V_{base;1}$ is the base voltage on primary side of the transforer, the base voltage on the secondary side, $V_{base;2}$ is then:

$$V_{base;2} = V_{base;1} \cdot \frac{N_2}{N_1} \tag{1.82}$$

Similarly:

$$I_{base;2} = I_{base;1} \cdot \frac{N_1}{N_2} \tag{1.83}$$

Assuming an effective impedance for the transformer referred to the primary side of the transformer of $Z_{trans;1}$ then this can be transferred to the secondary side as $Z_{trans;2}$ using the turns ratio squared as follows:

$$Z_{trans;2} = Z_{trans;1} \cdot \left(\frac{N_2}{N_1}\right)^2 \tag{1.84}$$

The per unit impedance of the transformer referred to the primary ($Z_{tpu;1}$) is therefore:

$$Z_{tpu;1} = Z_{trans;1} \cdot \frac{I_{base;1}}{V_{base;1}} \tag{1.85}$$

Similarly for $Z_{tpu;2}$:

$$Z_{tpu;2} = Z_{trans;2} \cdot \frac{I_{base;2}}{V_{base;2}} \tag{1.86}$$

Substituting equations (1.82), (1.83) and (1.84) in equation (1.86):

$$\begin{aligned} Z_{tpu;2} &= Z_{trans;1} \cdot \left(\frac{N_2}{N_1}\right)^2 . I_{base;1} \cdot \left(\frac{N_1}{N_2}\right) \cdot \frac{1}{V_{base;1}} \cdot \left(\frac{N_1}{N_2}\right) \\ &= Z_{trans;1} \cdot \frac{I_{base;1}}{V_{base;1}} = Z_{tpu;1} \end{aligned} \tag{1.87}$$

The per-unit impedance of a transformer is therefore the same irrespective of whether it is viewed from the primary or secondary side and a transformer can be represented on the per unit diagram simply by its impedance.

Comment

The per unit value of a transformer or machine is often expressed as a percentage value in which case

$$12\% \equiv 0.12 \text{ pu.}$$

Example 1.5 Per-unit values

Draw the per-unit diagram for the power system shown in Fig. 1.31 using a base value for voltage of 33 000 volts (line) at busbar 1 and a base value for volt-amperes of 15 MVA. Hence find the current in load 1 with an input voltage to transformer T_1 of 138 600 volts (line).

The bases values of voltage in the system are:

Transmission system	132 000
Busbar 1	33 000
Busbar 2	11 000

For transformer T_1 the per unit value is:

$$\boldsymbol{Z}_{\mathrm{T1;pu}} = 0.10\left(\frac{15}{15}\right) = 0.1\mathrm{j}\ \mathrm{pu}$$

since the base voltage is unchanged. Similarly for transformer T_2:

$$\boldsymbol{Z}_{\mathrm{T2;pu}} = 0.15\left(\frac{15}{2}\right) = 1.125\mathrm{j}\ \mathrm{pu}$$

Assuming that the loads can be represented as simple impedances, then for load L_3 and referring to Example 1.4:

$$\boldsymbol{Z}_{\mathrm{L3;pu}} = \left(\frac{15}{11^2}\right)(91.86 + 41.58\mathrm{j}) = 11.39 + 5.15\mathrm{j} = 12.5\underline{|24.33^\circ}\ \mathrm{pu}$$

Similarly, for loads L_1 and L_2:

$$\boldsymbol{Z}_{\mathrm{L1;pu}} = 1.29 + 0.765\mathrm{j} = 1.5\underline{|30.67^\circ}\ \mathrm{pu}$$

and

$$\boldsymbol{Z}_{\mathrm{L2;pu}} = 29.4 - 5.97\mathrm{j} = 30\underline{|-11.48^\circ}\ \mathrm{pu}$$

Also

$$V_{\mathrm{pu}} = \frac{138\ 600}{132\ 000} = 1.05\ \mathrm{pu}$$

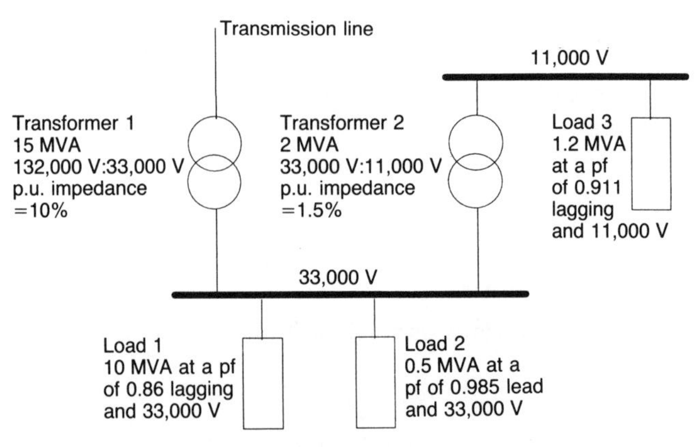

Fig. 1.31 Power system for Example 1.5.

The per-unit diagram is then as shown in Fig. 1.32.
Solving

$$I_{s:pu}=0.778\underline{|-32.49^\circ}\text{ pu}$$

then

$$I_{1:pu}=I_{s:pu}\left(\frac{Z_{2;pu}}{Z_{L1;pu}+Z_2}\right)=0.674\underline{|-34.35^\circ}\text{ pu}$$

where

$$Z_{2:pu}=Z_{L2:pu}\|(Z_{L3:pu}+Z_{T2:pu})$$

Now, at busbar 1

$$I_{base}=\frac{VA_{base}}{\sqrt{3}\,V_{base}}=\frac{15\times 10^6}{\sqrt{3}\times 33\,000}=262.4\text{ A}$$

Therefore

$$I_2=0.674\times 262.4=176.9\text{ A}$$

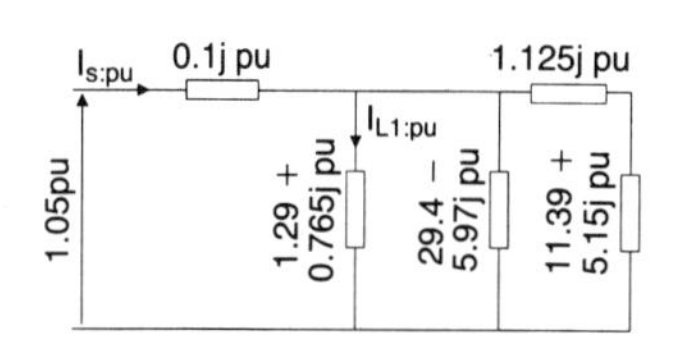

Fig. 1.32 Single line diagram for the circuit of Fig. 1.31.

1.4.6 *Unbalanced three-phase loads*

In an unbalanced three-phase load, the individual impedances making up the load have different magnitudes and phase angles.

(a) *Unbalanced four-wire star connected load*

In the unbalanced four-wire star of Fig. 1.33 the presence of the neutral conductor ensures that the phase voltages appear across the individual loads when:

$$\begin{vmatrix} V_1 \\ V_2 \\ V_3 \end{vmatrix}=\begin{vmatrix} I_1\,.\,Z_1 \\ I_2\,.\,Z_2 \\ I_3\,.\,Z_3 \end{vmatrix}=\begin{vmatrix} I_1\,.\,Z_1\underline{|\phi_1} \\ I_2\,.\,Z_2\underline{|\phi_2} \\ I_3\,.\,Z_3\underline{|\phi_3} \end{vmatrix} \tag{1.88}$$

and

$$I_n=I_1+I_2+I_3 \tag{1.89}$$

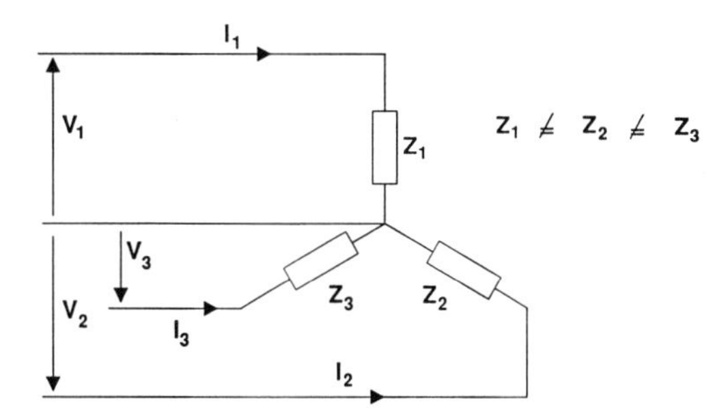

Fig. 1.33 Unbalanced four-wire star connected load.

The three-phase power supplied to the load is then:

$$P = V_1 . I_1 . \cos \phi_1 + V_2 . I_2 . \cos \phi_2 + V_3 . I_3 \cos \phi_3$$
$$= I_1^2 . R_1 + I_2^2 . R_2 + I_3^2 . R_3 \qquad (1.90)$$

where

$$R_1 = Z_1 . \cos \phi_1$$
$$R_2 = Z_2 . \cos \phi_2$$

and

$$R_3 = Z_3 . \cos \phi_3$$

(b) *Unbalanced three-wire star connected load*

The absence of the neutral conductor in the unbalanced three-wire star connected load of Fig. 1.34 means that the **star point** of the load is no longer at the same potential as the supply neutral and hence that the phase voltages no longer appear across the corresponding load impedances. The currents must therefore be calculated from the line voltages when:

$$\begin{vmatrix} \boldsymbol{V}_{12} \\ \boldsymbol{V}_{23} \\ \boldsymbol{V}_{31} \end{vmatrix} = \begin{vmatrix} \boldsymbol{I}_1 . \boldsymbol{Z}_1 - \boldsymbol{I}_2 . \boldsymbol{Z}_2 \\ \boldsymbol{I}_2 . \boldsymbol{Z}_2 - \boldsymbol{I}_3 . \boldsymbol{Z}_3 \\ \boldsymbol{I}_3 . \boldsymbol{Z}_3 - \boldsymbol{I}_1 . \boldsymbol{Z}_1 \end{vmatrix} \qquad (1.91)$$

Unfortunately, these equations are not mutually independent as any two can be used to generate the third. In order to solve for the phase (or line) currents the equation (1.90) for the currents at the star point must be used.

$$\boldsymbol{I}_1 + \boldsymbol{I}_2 + \boldsymbol{I}_3 = 0 \qquad (1.92)$$

The three-phase power supplied to the load is then:

$$P = I_1^2 . R_1 + I_2^2 . R_2 + I_3^2 . R_3 \qquad (1.93)$$

where

$$R_1 = Z_1 . \cos \phi_1$$
$$R_2 = Z_2 . \cos \phi_2$$

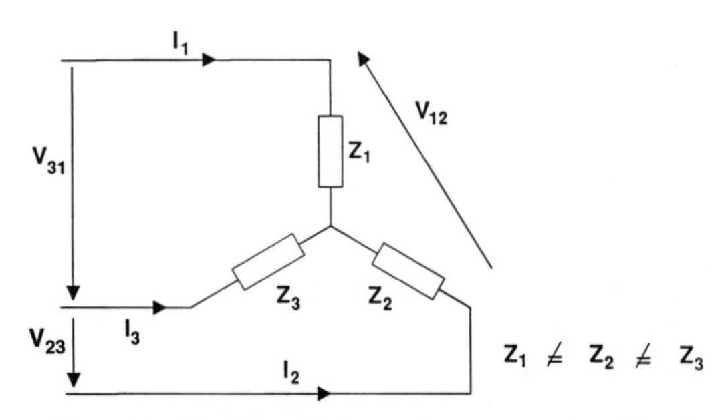

Fig. 1.34 Unbalanced three-wire star connected load.

and

$$R_3 = Z_3 \, . \cos \phi_3$$

The voltage of the star point with respect to the neutral point of the balanced three-phase supply voltages is then:

$$\boldsymbol{V}_n = \boldsymbol{V}_1 - \boldsymbol{I}_1 \, . \, \boldsymbol{Z}_1 = \boldsymbol{V}_2 - \boldsymbol{I}_2 \, . \, \boldsymbol{Z}_2 = \boldsymbol{V}_3 - \boldsymbol{I}_3 \, . \, \boldsymbol{Z}_3 \tag{1.94}$$

Example 1.6 Unbalanced three-wire star connected load
Figure 1.35 shows an unbalanced star connected load connected to a balanced, three-phase 660 volt (line) supply. Calculate the current in each phase of the load and the total power supplied to the load.

Now, with $\boldsymbol{V}_{12}$ as reference:

$$660 + 0\text{j} = \boldsymbol{I}_1(8 + 6\text{j}) - \boldsymbol{I}_2(10 + 0\text{j}) \tag{i}$$

$$660\underline{|-120^\circ} = -330 - 571.6\text{j} = \boldsymbol{I}_2(10 + 0\text{j}) - \boldsymbol{I}_3(7.07 + 7.07\text{j}) \tag{ii}$$

and

$$\boldsymbol{I}_1 + \boldsymbol{I}_2 + \boldsymbol{I}_3 = 0 \tag{iii}$$

Therefore, from (ii) and (iii):

$$-330 - 571.6\text{j} = \boldsymbol{I}_2(17.07 + 7.07\text{j}) + \boldsymbol{I}_1(7.07 + 7.07\text{j}) \tag{iv}$$

Then, multiplying (i) by (17.07 + 7.07j)

$$11\,266 + 4666\text{j} = \boldsymbol{I}_1(94.15 + 159\text{j}) - \boldsymbol{I}_2(170.7 + 70.7\text{j}) \tag{v}$$

and multiplying (iv) by 10 + 0j

$$-3300 - 5716\text{j} = \boldsymbol{I}_2(170.7 + 70.7\text{j}) + \boldsymbol{I}_1(70.7 + 70.7\text{j}) \tag{vi}$$

Combining (v) and (vi)

$$7966 - 1049.6\text{j} = \boldsymbol{I}_1(164.9 + 229.7\text{j}) = \boldsymbol{I}_1 \times 228\underline{|55.33^\circ}$$

Therefore

$$\boldsymbol{I}_1 = 12.96 - 25.27\text{j} = 28.4\underline{|-62.84^\circ} \text{ A}$$

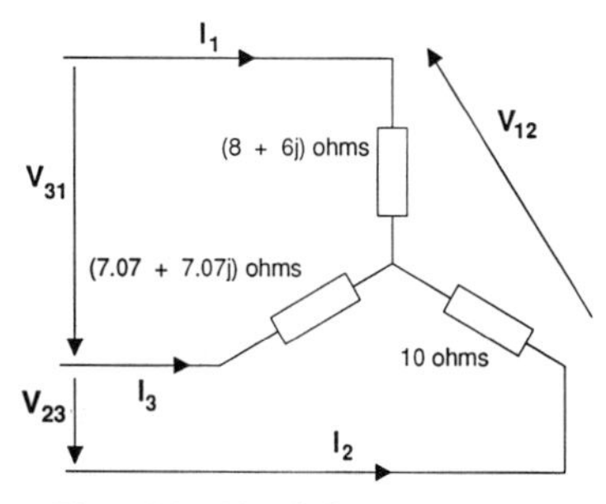

Fig. 1.35 Circuit for Example 1.6.

Substituting for I_1 in (i):

$$I_2 = 2.84\underline{|-62.84^\circ}\ (8+6j) - 66 = -40.47 - 12.44j = 42.34\underline{|-162.9^\circ}\ \text{A}$$

Then

$$I_3 = -(I_1 + I_2) = 27.51 + 37.71j = 46.68\underline{|53.89^\circ}\ \text{A}$$

Check using

$$660\underline{|-240^\circ} - (I_3 Z_3 - I_1 Z_1) = -2.86 - 13.97j = 14.26\underline{|-101.6^\circ}\ \text{A}$$

This is an error in magnitude of 2.29% which is well within that to be expected allowing for rounding errors in the calculation.

Power supplied is then:

$$\text{Power} = I_1^2 R_1 + I_2^2 R_2 + I_3^2 R_3 = 28.4^2 \times 8 + 42.34^2 \times 10 + 46.68^2 \times 7.07$$
$$= 39\ 780\ \text{watts}$$

(c) *Unbalanced delta connected load*

In the unbalanced delta connected load of Fig. 1.36 a line voltage appears across each of the load impedances when:

$$\begin{vmatrix} V_{12} \\ V_{23} \\ V_{31} \end{vmatrix} = \begin{vmatrix} I_{12} \cdot Z_{12} \\ I_{23} \cdot Z_{23} \\ I_{31} \cdot Z_{31} \end{vmatrix} \tag{1.95}$$

and

$$\begin{vmatrix} I_1 \\ I_2 \\ I_3 \end{vmatrix} = \begin{vmatrix} I_{12} - I_{31} \\ I_{23} - I_{12} \\ I_{31} - I_{23} \end{vmatrix} \tag{1.96}$$

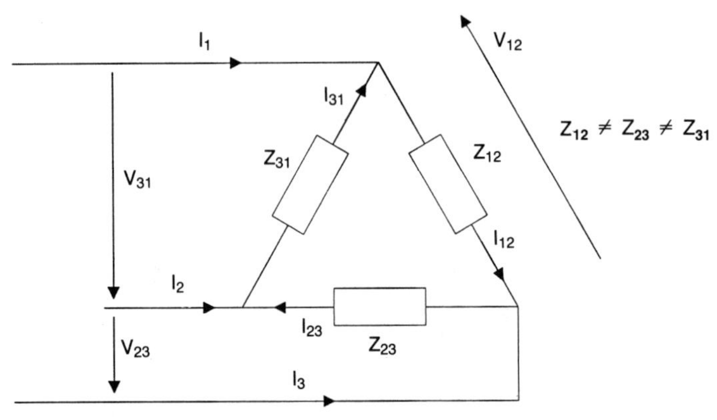

Fig. 1.36 Unbalanced delta connected load.

The three-phase power supplied is:

$$P = V_{12} . I_{12} . \cos \phi_{12} + V_{23} . I_{23} . \cos \phi_{23} + V_{31} . I_{31} . \cos \phi_{31}$$
$$= I_{12}^2 . R_{12} + I_{23}^2 . R_{23} + I_{31}^2 . R_{31} \quad (1.97)$$

where

$$R_{12} = Z_{12} . \cos \phi_{12}$$
$$R_{23} = Z_{23} . \cos \phi_{23}$$

and

$$R_{31} = Z_{31} . \cos \phi_{31} .$$

Example 1.7 Unbalanced delta connected load
In Fig. 1.37 an unbalanced delta connected load is connected to a balanced three-phase supply of 660 volts (line). Find the current in each leg of the delta, the associated line currents and the power in the load.

Taking V_{12} as reference:

$$\boldsymbol{I}_{12} = \frac{660\underline{|0^\circ}}{\boldsymbol{Z}_{12}} = 66\underline{|-36.87^\circ} = 52.8 - 39.6\mathrm{j} \text{ A}$$

$$\boldsymbol{I}_{23} = \frac{660\underline{|-120^\circ}}{\boldsymbol{Z}_{23}} = 66\underline{|-120^\circ} = -33 - 57.2\mathrm{j} \text{ A}$$

$$\boldsymbol{I}_{31} = \frac{660\underline{|-240^\circ}}{\boldsymbol{Z}_{31}} = 66\underline{|-285^\circ} = 17.1 + 63.75\mathrm{j} \text{ A}$$

Then

$$\boldsymbol{I}_1 = \boldsymbol{I}_{12} - \boldsymbol{I}_{31} = 35.7 - 103.35\mathrm{j} = 109.34\underline{|-70.94^\circ} \text{ A}$$
$$\boldsymbol{I}_2 = \boldsymbol{I}_{23} - \boldsymbol{I}_{12} = -85.4 - 17.6\mathrm{j} = 87.19\underline{|-168.35^\circ} \text{ A}$$
$$\boldsymbol{I}_3 = \boldsymbol{I}_{31} - \boldsymbol{I}_{23} = 50.1 + 120.95\mathrm{j} = 130.92\underline{|67.5^\circ} \text{ A}$$

Power $= I_{12}^2 R_{12} + I_{23}^2 R_{23} + I_{31}^2 R_{31} = 66^2 \times 8 + 66^2 \times 10 + 66^2 \times 7.07 = 109\ 200$ watts.

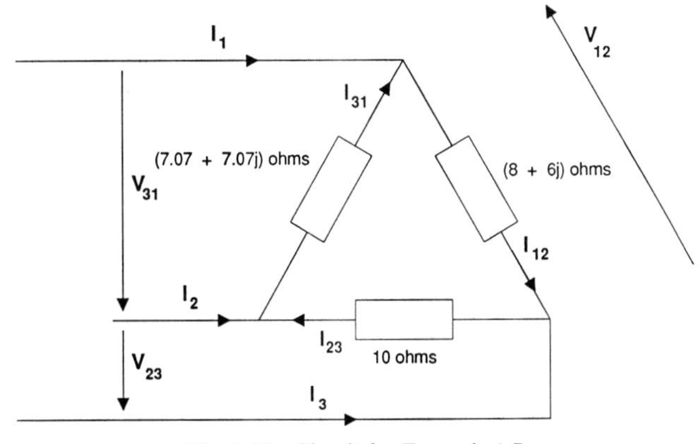

Fig. 1.37 Circuit for Example 1.7.

Exercises

1 For the single-phase a.c. circuit of Fig. 1.38 find the current drawn from the single-phase, 240 volt a.c. supply, the power delivered and the power factor.

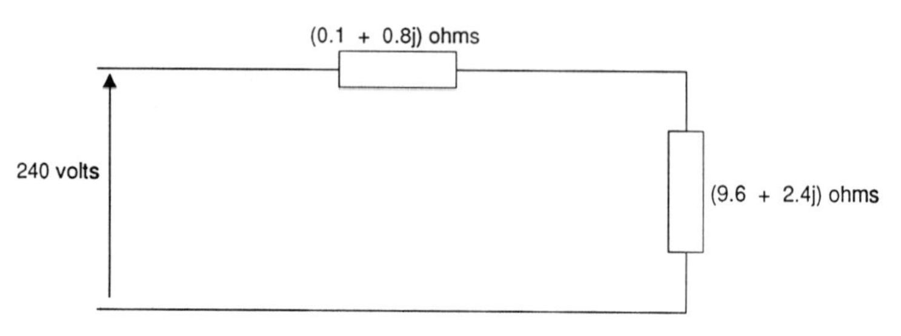

Fig. 1.38 Circuit for Exercise 1.

2 A 100 kVA load with a power factor of 0.92 lagging is connected to a single-phase 6350 volt supply. What value of capacitor, expressed as an appropriate VAR value, would be required to produce an effective power factor of unity at the load terminals?

3 If the load of Exercise 2 is now reduced to 60 kVA at a power factor of 0.9 while the value of power factor correction capacitor remains unaltered, what would be the new power factor at the load terminals, the volt-amperes supplied by the source (S_s) and the current drawn from the source?

4 For the circuit of Fig. 1.39 $Z_Y = 10 + 0j\ \Omega$ and $Z_\Delta = 20 + 10j\ \Omega$ and it is supplied from a balanced, three-phase, 220 volt (line) source. Determine the line currents drawn from the supply for this combination and find the per-phase impedance of the star connected load that would result in the same line currents.

5 Using per-unit values calculate the current drawn from the 11 kV source in the system of Fig. 1.40. Use a base value for voltage of 415 volts (line) at busbar 1 and a base value for volt-amperes of 1 MVA.

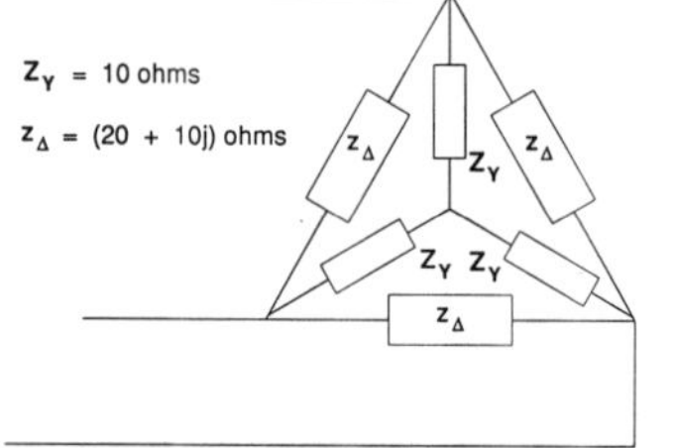

Fig. 1.39 Load for Exercise 4.

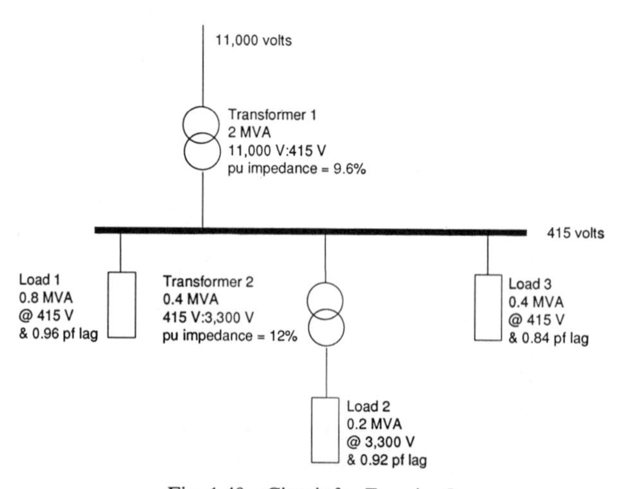

Fig. 1.40 Circuit for Exercise 5.

6 (a) An unbalanced, three-wire star connected load is made up of three impedances as follows:

$Z_1 = 10 + 0j\ \Omega$

$Z_2 = 5 + 5j\ \Omega$

and

$Z_3 = 10 + 5j\ \Omega$

and is connected to a balanced, three-phase, 415 volt (line) supply. Calculate the current in each line of the supply.

(b) If a neutral conductor is now added to convert the load into a four-wire star, what will be the new values for the line currents and what will be the current in the neutral conductor?

7 An unbalanced, three-wire star connected load is made up of three impedances as follows:

$Z_1 = 10 + 0j\ \Omega$

$Z_2 = 10 + 5j\ \Omega$

and

$Z_3 = 10 - 5j\ \Omega$

and is connected to a balanced, three-phase, 100 volt (line) supply. Find the current in each leg of the load and the total power dissipated by the load.

2 Power electronics

Solid state devices for the control of electrical power are now available with ratings from a few watts to hundreds of kilowatts and are used for a wide variety of applications including:

- lighting control
- switch mode power supplies
- uninterruptable power supplies
- voltage regulation
- a.c. motor control
- d.c. motor control
- power transmission
- heating control

In practice, power electronic devices are used as switches and may be classified according to their operation into three groups:

- Devices whose turn-on and turn-off are determined entirely by circuit conditions. The **diode** is the only device in this category.
- Devices whose turn-on can be controlled by an external signal but whose turn-off is determined by circuit conditions. **Thyristors** and **triacs** are in this category.
- Devices whose turn-on and turn-off can be controlled by an external signal. **Gate turn-off** (GTO) **thyristors**, **power transistors**, **insulated gate bipolar transistors** and **power MOSFETs** all come within this category of **self-commutated** device.

Figure 2.1 provides some details of the common power electronic devices along with their circuit representation.

2.1 Power electronic devices

2.1.1 *Uncontrolled devices*

(a) *Diodes*

The diode is the simplest of the power electronic devices. An ideal diode will turn on when it becomes **forward biased** and will then continue to conduct until the forward current falls to zero. In practice, a forward voltage of around 0.7 volts is required to turn the diode on and a **forward voltage drop** of the same magnitude is then maintained across the device while it is conducting.

Comment

The diode is forward biased when terminal I in Fig. 2.1 is at a higher positive potential than terminal II, in which case the voltage v_d is greater than zero.

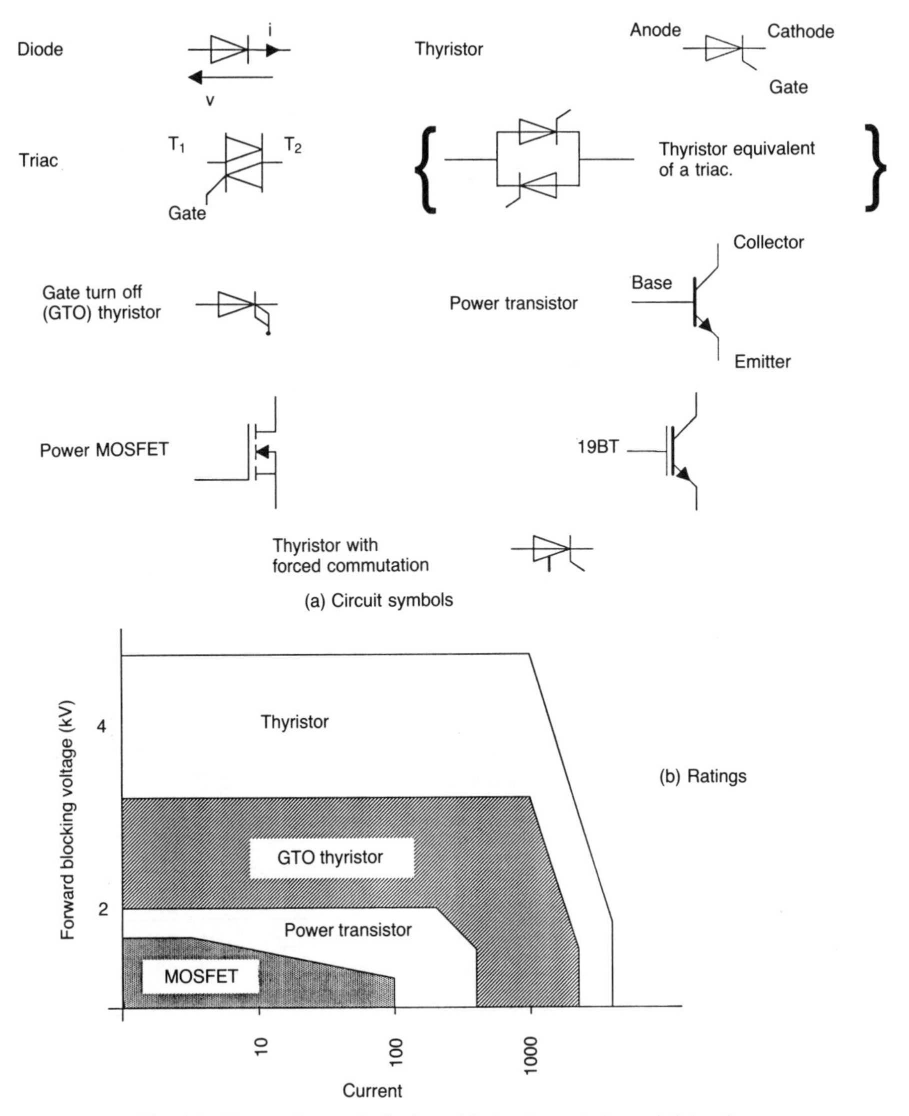

Fig. 2.1 Power electronic devices: (a) circuit symbols; and (b) ratings.

2.1.2 *Devices with controlled turn-on*

(a) *Thyristors*

A thyristor, like a diode, will conduct in one direction only but is capable of blocking conduction in either direction and, unlike a diode, does not automatically turn on when it becomes forward biased. Turn-on or **firing** of the thyristor when in a forward biased condition is achieved by means of the injection of a small current into the **gate circuit**. This **gate current** must be maintained until the forward current through the thyristor has reached the **latching current** for the device; after this point conduction is self-sustaining and the gate current can be removed. In practice, however, the gate current may well be maintained during the whole of the conduction period or a **pulse train** used to supply the gate to ensure proper firing under all conditions. This is particularly the case where the thyristor is being used to supply an inductive load.

The characteristics of a thyristor gate circuit are shown in Fig. 2.2 in which the **area of certain firing** is bounded by:

- the upper limit of gate resistance (I_g against V_g);
- the lower limit of gate resistance (I_g against V_g);
- the minimum values of I_g and V_g for triggering (these vary with temperature);
- the gate power limit.

The combination of gate voltage (V_g) and gate current to be used is then obtained by reference to the gate load line when the actual operating point is defined by the intersection with the gate resistance line for the particular thyristor.

Comment

Referring to Fig. 2.2, the gate load load line is drawn between V and V/R_g on the voltage and current axes respectively and has a slope of $-R_g$.

In many applications all the terminals of the thyristor are at some potential with respect to ground in which case it is necessary to isolate the gate circuit from the thyristor using circuit arrangements such as those shown in Fig. 2.3.

Comment

The diode shown in Fig. 2.3(a) across the transformer primary short circuits the current pulse that results when the transistor is turned off. Without this diode the rapid collapse of primary current, and hence of transformer core flux would result in the production of

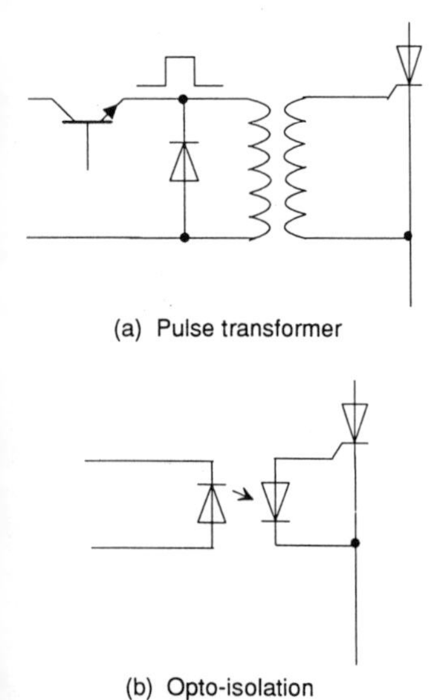

Fig. 2.3 Gate circuit isolation: (a) pulse transformation; and (b) opto-isolation.

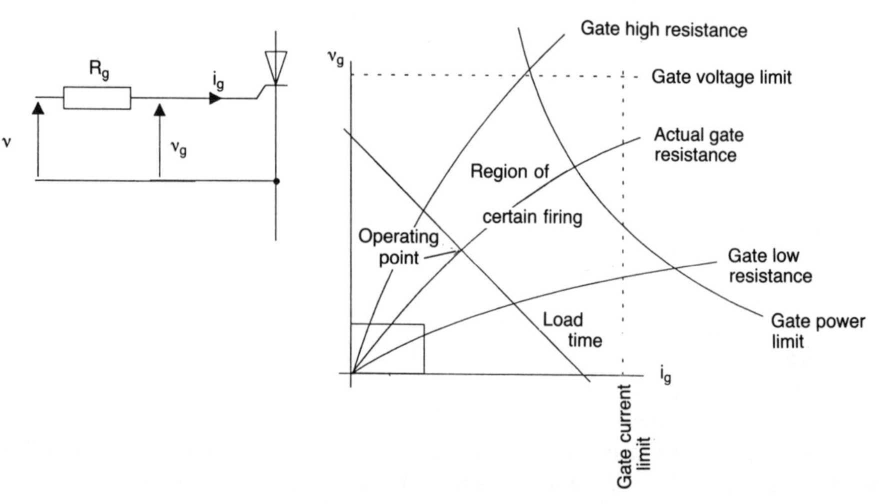

Fig. 2.2 Thyristor gate circuit characteristics.

large e.m.f.s. in the primary and secondary winding with possible damage to either or both of the transistor and thyristor.

Once on, the thyristor will continue to conduct until the forward current falls below the level referred to as the **holding current** at which point the thyristor begins to turn off. Once the thyristor has been turned off, a reverse voltage must be applied for a short period in order to re-establish the internal conditions for forward blocking

In many practical applications, the conditions for thyristor turn-off occur automatically as a result of the circuit configuration. These are referred to as **naturally commutated** circuits. In other applications, particularly where the switching of a d.c. source voltage is involved, additional, external circuitry must be used to generate the required conditions to turn the thyristor off. This is referred to as **forced commutation**.

Comment

The turn-on and turn-off of a thyristor are complex physical processes and performance is influenced by factors such as the load inductance, duty cycle and temperature.

Thyristors have the highest ratings among power electronic devices and this combined with the relatively low turn-on power requirement means that they are the primary choice for applications such as high-power variable d.c. voltage supplies and high-voltage d.c. (HVDC) power transmission. Where controlled turn-off is required, the combination of thyristors with forced commutation has largely been superseded by the self-commutated devices.

Example 2.1 Thyristor turn-on

A thyristor with a latching current of 40 mA is connected as in Fig. 2.4 to supply a load of 15 ohms and 0.4 henries. If the gate signal is applied to the thyristor at the instant of maximum supply voltage, what will be the minimum duration of that gate signal in order to ensure turn-on of the thyristor?

Following application of the gate signal

$$100\cos\omega t = i\,.\,R + L\,.\frac{\mathrm{d}i}{\mathrm{d}t}$$

Solving for current:

$$i = 100\,\frac{[\cos(\omega t - \phi) - \cos\phi\,.\,\mathrm{e}^{-Rt/L}]}{(R^2 + \omega^2 L^2)^{\frac{1}{2}}}$$

in which

$$\phi = \tan^{-1}\left(\frac{\omega L}{R}\right) = 83.19° = 1.452 \text{ radian}$$

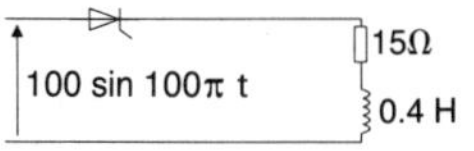

100 sin 100π t
15Ω
0.4 H

Fig. 2.4 Example 2.1: thyristor switching.

and

$$(R^2+\omega^2L^2)^{\frac{1}{2}}=126.6$$

Therefore

$$i=0.79[\cos(\omega t-1.452)-0.1186\,e^{-37.6t}]$$

Solving graphically gives a value for t of 161 μs.

(b) *Triacs*

A triac acts like a pair of inverse parallel thyristors to allow conduction in either direction. Triacs have lower ratings than thyristors but have found extensive application in the control of domestic lighting systems and in domestic products.

For best performance a positive gate signal – current in to the gate – should be used to turn on a triac when terminal T_1 in Fig. 2.1 is positive with respect to terminal T_2 and a negative gate current – current out of the gate – when T_2 is positive with respect to T_1. If a single direction of gate current is to be used then this should be in the positive sense.

2.1.3 *Self-commutated devices*

(a) *Gate turn-off thyristors*

The gate turn-off or GTO thyristor functions during turn-on like a conventional thyristor but can be turned off by the injection of a reverse current, typically 25% or more of the forward current, into the gate of the device. GTO thyristors offer relatively high ratings with a good switching performance and have been applied to systems such as **pulse width modulated (PWM) inverters** for induction motor control.

A disadvantage of the GTO thyristor is that its reverse voltage capability is much less than that of a conventional thyristor and for this reason it would typically be used in series with a diode to provide the necessary capability.

(b) *Power transistors*

Power transistors allow control of both turn-on and turn-off performance for switching purposes. Their primary disadvantage is the need to maintain a continuous base current, which would typically be of the order of 10% to 20% of the forward current, while the device is conducting. The use of integral base drive circuits such as the Darlington circuit of Fig. 2.5 on the same chip reduces the size of the external base current required but at the expense of a reduced switching performance of the power device.

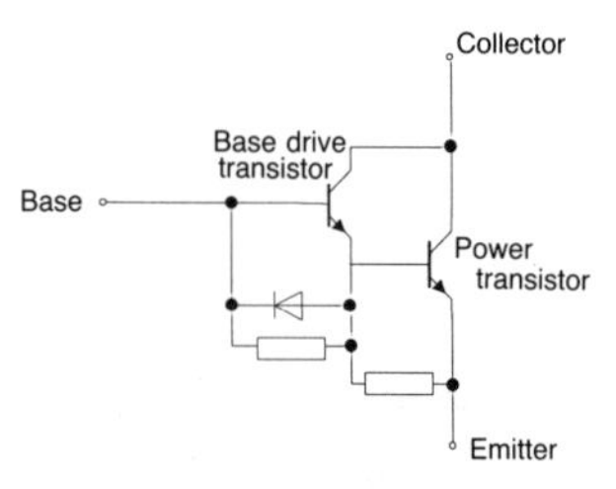

Fig. 2.5 Practical Darlington circuit.

(c) *Power MOSFETs*

Power MOSFETs are developments of conventional MOS (metal oxide silicon) technology whose 'on' or 'off' state is determined by the application of a voltage to the gate terminal shown in Fig. 2.1, the main current then flows between the drain and source. Power MOSFETs have high gate impedances, reducing the gate current and hence the gatepower levels required for turn-on and to maintain conduction. This low gate current requirement means that it is often possible to drive a power MOSFET directly from standard logic, VLSI or microcomputer ports. As the gate circuit is primarily capacitive the gate current source must be properly matched, particularly where high-speed operation is required.

A reverse diode is typically incorporated into the construction of a power MOSFET and a series diode must be used where it is required to block any reverse conduction.

(d) *Insulated gate bipolar transistors*

The insulated gate bipolar transistor (IGBT) combines a MOSFET type gate with a bipolar switching element to give a device which has a good switching performance with reduced gate current levels. Because of their overall performance capabilities, IGBTs are displacing conventional power transistors for many applications in their power range.

2.2 Device characteristics

Power electronics devices are principally rated in terms of their continuous, pulsed and r.m.s. current capacities and their ability to withstand both forward and reverse voltages. Secondary characteristics include the maximum rates of change of current (di/dt) and voltage (dv/dt) that the device can withstand on turn-on and turn-off and the $\int i^2 dt$ rating relating to the maximum permitted temperature rise in the device under transient conditions. The dv/dt values of the device are typically controlled by **snubber circuits**, examples of which are shown in Fig. 2.6. The $\int i^2 dt$ rating determines the choice of fuse used.

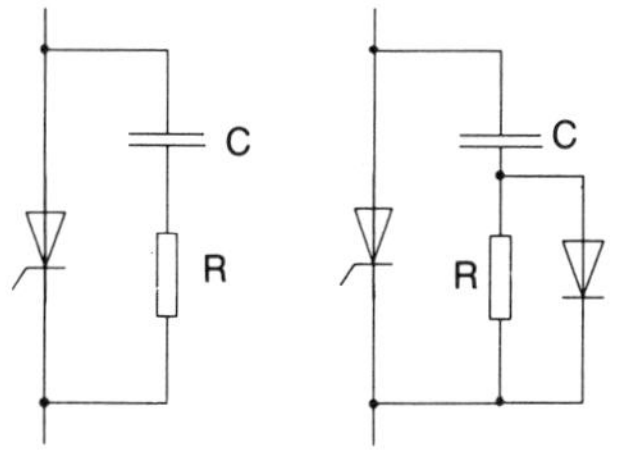

Fig. 2.6 Simple snubber circuits.

Comment

The design of snubber circuits is a complex process involving the relationships between the components of the snubber circuit and the performance characteristics of the device. Computer programs are now available to assist in the optimization of snubber design.

2.2.1 *Heat transfer and cooling*

The major sources of loss in a power electronic switching device are:

- on-state losses resulting from the combination of the current through the device and the device forward volt drop;
- off-state losses resulting from the device leakage current and the voltage across the device;
- switching losses incurred during the transition from the on-state to the off-state and *vice versa*. These may well be the major source of loss in applications such as pulse-width modulated (PWM) inverter drives.

These losses result in the internal heating of the device and heat energy must be removed via the path:

1. from the internal device junction to the case;
2. from the case to a heat transfer system such as a fin;
3. from the heat transfer system to an ambient temperature heat sink.

The effect of the various stages of the heat removal process can be modelled under steady state conditions by the heat transfer circuit of Fig. 2.7 using thermal resistors when:

$$\theta_j = \theta_a + P.(R_{jc} + R_{ch} + R_{ha}) \tag{2.1}$$

where:

θ_j is the junction temperature;
θ_a is the ambient temperature;
P is the internal power developed as loss in the device;
R_{jc} is the effective steady state thermal resistance between the device junction and the case of the device;
R_{ch} is the effective steady state thermal resistance between the case of the device and the heat transfer system;
R_{ha} is the effective steady state thermal resistance between the heat transfer system and ambient.

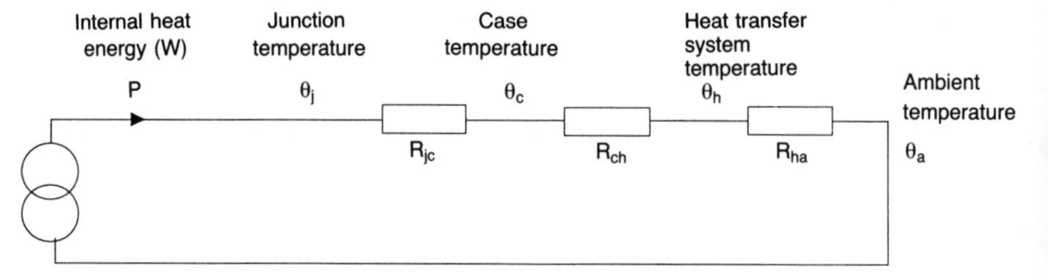

Fig. 2.7 Steady state heat transfer circuit.

Example 2.2 Heat transfer
A thyristor has a thermal resistance of 0.74 °C/W between its junction and the hea transfer system (Note: This is the combination of R_{jc} and R_{jh} in equation (2.1)) anc 2.04 °C/W between the heat transfer system and ambient. What will be the junctio temperature for an ambient temperature of 25 °C when 36 watts of thermal power ar being produced in the thyristor?

$$\theta_j = 25 + 36(0.74 + 2.04) = 125.08\ °C$$

During transient operation such as would occur during pulsed operation, overload or faults, the temperature rise can be estimated by using the transient therma impedance which takes into account the effect of the thermal mass of the device i delaying the transfer of heat energy from the junction. The transient therma

input over a defined interval such that for a sudden increase in power dissipation at time $t=0$ from 0 to P_{th}:

$$P_{th} \cdot Z_{th}(t) = \delta\theta(t) \quad (2.2)$$

where:

$Z_{th}(t)$ is the transient thermal impedance after time t;
$\delta\theta(t)$ is the temperature rise after time t.

Consider a power semiconductor providing a regular train of current pulses as in Fig. 2.8. The mean junction temperature can then be found from the mean power loss and the transient thermal impedance at $t=\infty$, namely R_{ja}, as:

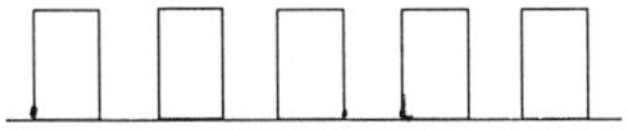

Fig. 2.8 Pulse train.

$$\theta_j = \theta_a + P_{mean} \cdot Z_{th}(\infty) = \theta_a + P_{mean} \cdot R_{ja} \quad (2.3)$$

where P_{mean} is the mean power dissipation and equals $\hat{P}(t/T)$.

The actual junction temperature varies about this mean value by an amount $\delta\theta_j$, a good approximation to which can be obtained in terms of the mean power as:

$$\delta\theta_j = Z_{th}(t_3)[\hat{P} - P_{mean}] - \hat{P} \cdot [Z_{th}(t_2) - Z_{th}(t_1)] \quad (2.4)$$

Figure 2.9 shows a typical curve of transient thermal impedance.

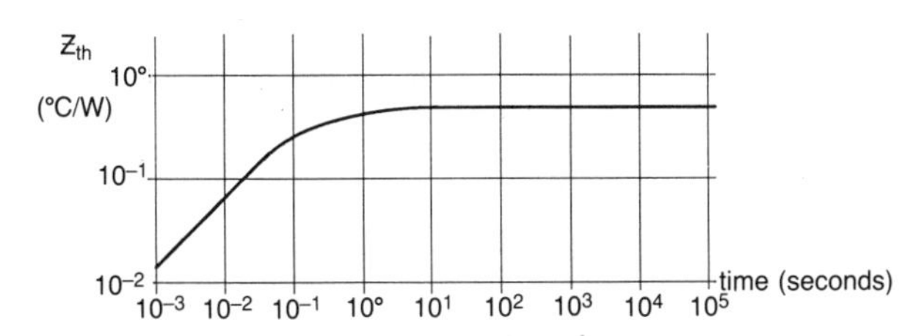

Fig. 2.9 Transient thermal impedance characteristic.

2.3 Naturally commutated converters

The function of a converter is that of transferring energy from an a.c. system to a d.c. system (**rectification**) and, in certain cases, from a d.c. system to an a.c. system (**inversion**). In a naturally commutated converter using a combination of diodes and thyristors the turn-off of the conducting devices takes place as a result of the variation of the a.c. voltage and no external, or forced commutation, circuitry is required. Types of naturally commutated converter are:

- Uncontrolled converter or rectifier, which uses diodes only. Output voltage is determined solely by the magnitude of the a.c. supply voltage. Can only transfer energy from the a.c. system to the d.c. system.
- Half-controlled converter, which uses a combination of diodes and thyristors. Output voltage is determined by the magnitude of the a.c. supply voltage and the firing angle of the thyristors. Can only transfer energy from the a.c. system to the d.c. system.
- Fully controlled converter, which uses thyristors throughout. Output voltage is determined by the magnitude of the a.c. supply voltage and the firing angle of the thyristors. Can transfer energy either from the a.c. system to the d.c. system or *vice versa*.

2.3.1 *A thyristor with an inductive load*

Figure 2.10(a) shows a single thyristor supplying an inductive load. As the thyristor turns off when the forward current falls below the holding level the effect of the load inductance will be to cause the thyristor to conduct beyond the voltage zero as in Fig. 2.10(b) before turning off.

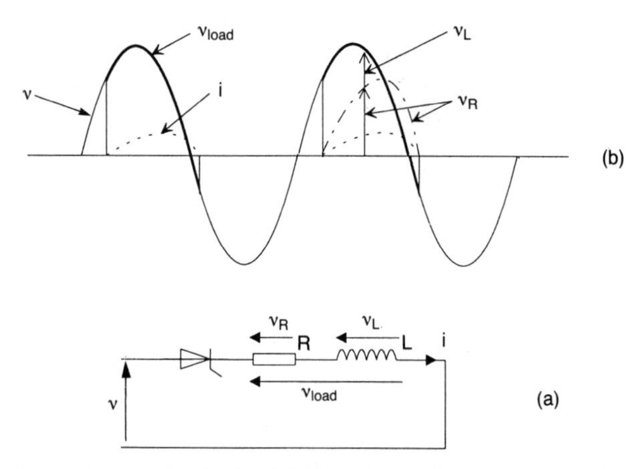

Fig. 2.10 Thyristor with an inductive load: (a) circuit; and (b) current and voltage waveforms.

2.3.2 *Uncontrolled converters*

Figure 2.11(a) shows the circuit configuration of a three-phase, uncontrolled half-bridge converter. Assuming an inductive load capable of maintaining a constant load current, the transfer of load from one diode to the next in sequence occurs at the crossover points of the voltage waveform. This is indicated in Fig. 2.11(b) assuming an instantaneous transfer of current. The magnitude of the output voltage is then:

$$V_0 = \frac{3}{2\pi}\int_{\frac{\pi}{6}}^{\frac{5\pi}{6}} \hat{V}\sin(\omega t)\mathrm{d}\omega t = \frac{3\sqrt{3}}{2\pi}\hat{V} \quad (2.5)$$

where $\hat{V}$ is the peak value of the applied voltage.

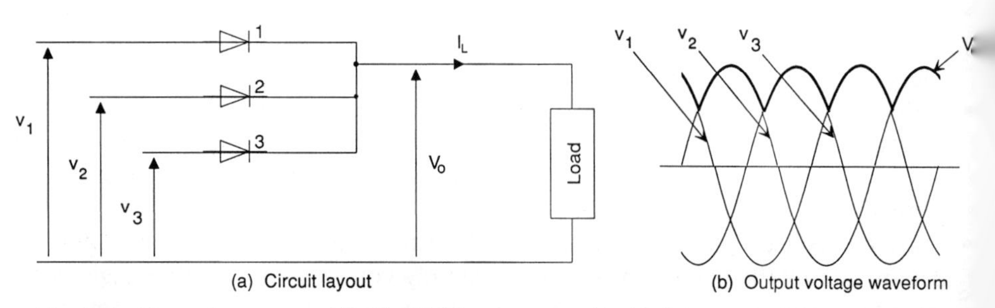

Fig. 2.11 Three-phase uncontrolled half-bridge converter: (a) circuit layout; and (b) output voltage waveform.

Comment

An uncontrolled converter is sometimes referred to simply as a rectifier.

2.3.3 *Fully controlled converters*

Figure 2.12(a) shows a three-phase, fully controlled half-bridge converter. Assuming as before that the load current is constant, then, for an instantaneous transfer of load current, the output voltage waveform has the form shown in Fig. 2.12(b). The magnitude of the output voltage in these circumstances is:

$$V_0 = \frac{3}{2\pi}\int_{\alpha+\frac{\pi}{6}}^{\alpha+\frac{5\pi}{6}} \hat{V}\sin(\omega t)\mathrm{d}\omega t = \frac{3\sqrt{3}}{2\pi}\hat{V}\cos\alpha \tag{2.6}$$

where α is the firing angle of the thyristors.

Comment

The difference between the three-phase, uncontrolled half-bridge converter and the three-phase, fully controlled half-bridge converter is that the diodes in the former have been replaced by thyristors in the latter.

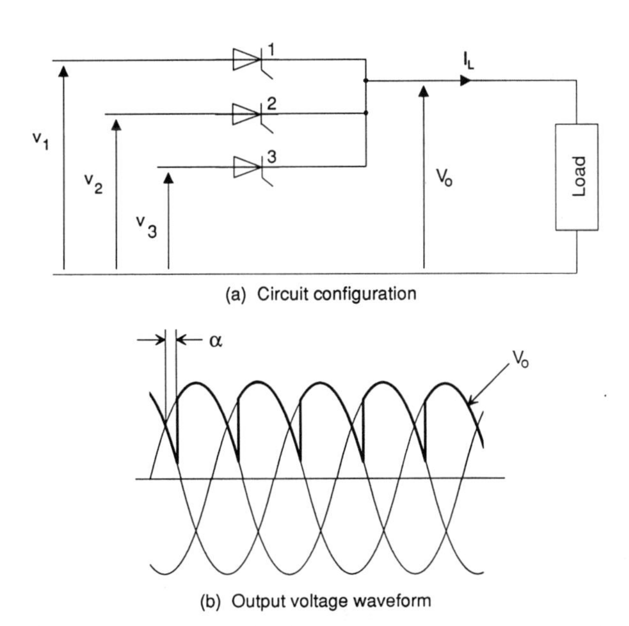

Fig. 2.12 Three-phase fully controlled half-bridge converter: (a) circuit configuration; and (b) output voltage waveform.

In Fig. 2.12(b) and equation (2.6) the firing angle α is measured from the point on the voltage waveform at which the particular thyristor becomes forward biased. This is also the point where the equivalent diode starts to conduct in the three-phase, half-bridge of Fig. 2.11.

(a) *Pulse number*

The operation of a converter may be expressed in terms of its pulse number (p). This is defined as the number of transitions between devices that occur during one cycle of the fundamental of the supply voltage waveform. Both the three-phase, uncontrolled half-bridge converter of Fig. 2.11(a) and the three-phase, fully controlled half-bridge converter of Fig. 2.12(a) have a pulse number of 3. The output voltage of an ideal converter when rectifying can be expressed in terms of its pulse number as:

$$V_{0;r} = \frac{p}{\pi}\hat{V}\sin(\pi/p)\,.\cos\alpha \tag{2.7}$$

Figure 2.13 gives examples of circuits of differing pulse numbers.

Comment

The output voltage of an uncontrolled p-pulse converter (rectifier) can be found by putting $\alpha = 0$ in equation (2.7).

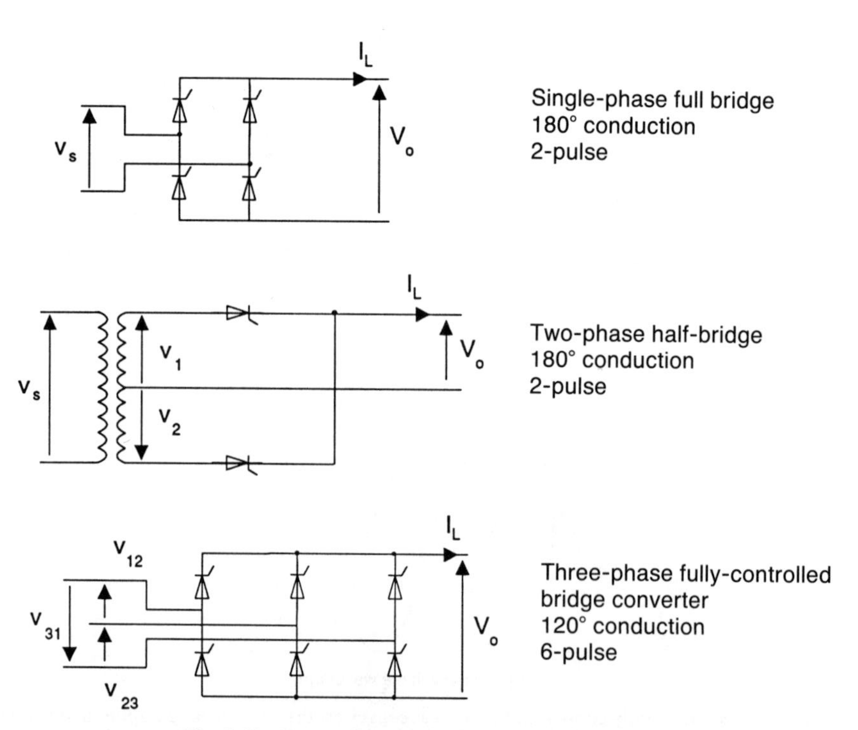

Fig. 2.13 Two-pulse and six-pulse converter circuits.

(b) *Overlap*

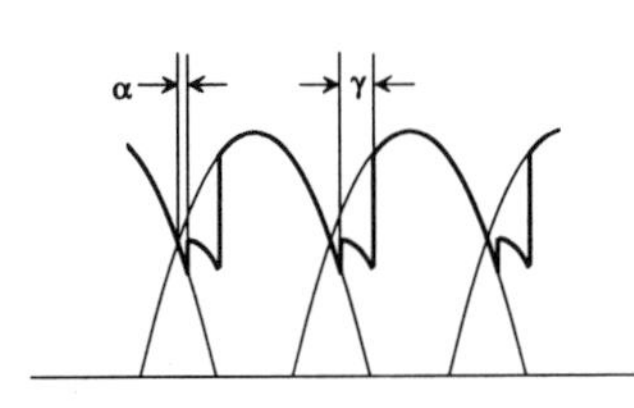

Fig. 2.14 Overlap.

In practice, current does not transfer instantaneously from the outgoing to the oncoming thyristor and there is a period during which both are conducting together. During this period, the load voltage is the mean of the voltages applied to the two thyristors as shown in Fig. 2.14. This condition is referred to as **overlap** and is measured by the **overlap angle** γ.

The effect of overlap is to reduce the mean output voltage when rectifying by an amount dependent upon the load current (I_L), the inductance of the supply (L) and the pulse number (p) of the converter according to the relationship:

$$V_{0;\text{rmean}} = \frac{\text{p}}{2\pi}\hat{V}\sin(\pi/\text{p})[\cos\alpha + \cos(\alpha+\gamma)]$$

$$= \frac{\text{p}}{\pi}\hat{V}\sin(\pi/\text{p}) . \cos\alpha - \frac{\text{p}\omega L}{2\pi} . I_L \quad (2.8)$$

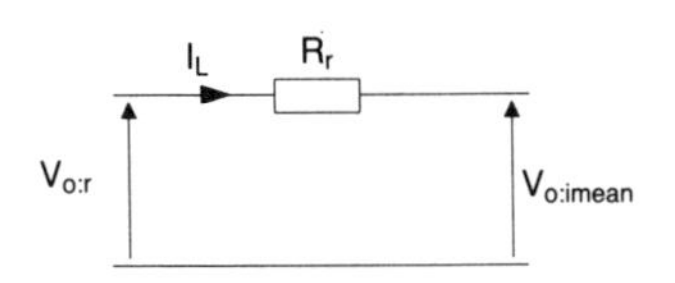

Fig. 2.15 Effect of overlap on rectifier performance.

The ($\text{p}\omega L/2\pi$) term in equation (2.8) affects the conditions on the d.c. side of the converter and is therefore dimensionally a resistance. Hence equation (2.8) can be expressed in the form:

$$V_{0;\text{rmean}} = V_{0;\text{r}} - R_r . I_L \quad (2.9)$$

The relationship between $V_{0;\text{rmean}}$ and $V_{0;\text{r}}$ is shown in Fig. 2.15.

Comment

The term $R_r . I_L$ in equation (2.9) represents a d.c. volt drop only, it does not have any associated power loss component.

(c) *Inversion*

If the firing angle of a fully controlled converter is made greater the 90° $\left(\frac{\pi}{2}\right)$ the output voltage will become negative as shown by Fig. 2.16. As the direction of current through the bridge cannot reverse, energy is now transferred from the d.c. system to the a.c. system and the bridge is said to be operating in inverting mode. The voltage equations for the converter in inverting mode are then, ignoring overlap:

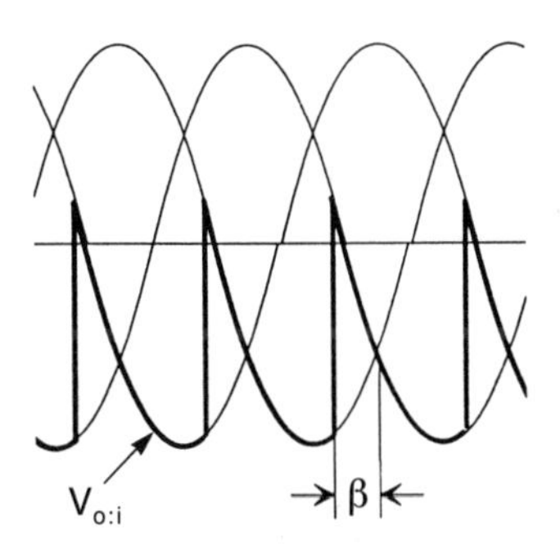

Fig. 2.16 Inversion.

$$V_{0;\text{i}} = \frac{\text{p}}{\pi}\hat{V}\sin(\pi/\text{p}) . \cos\beta \quad (2.10)$$

with overlap:

$$V_{0;\text{imean}} = \frac{\text{p}}{2\pi}\hat{V}\sin(\pi/\text{p})[\cos\beta + \cos(\beta-\gamma)]$$

$$= \frac{\text{p}}{\pi}\hat{V}\sin(\pi/p) . \cos\beta + \frac{\text{p}\omega L}{2\pi} . I_L = V_{0;\text{i}} + R_i . I_L \quad (2.11)$$

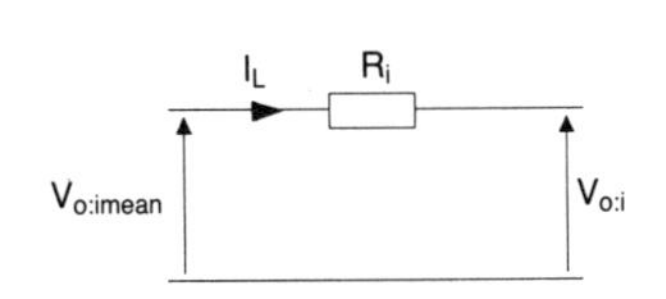

Fig. 2.17 Effect of overlap on inverter performance.

in which $\beta = \pi - \alpha$ and is referred to as the **firing advance angle**. The relationship between $V_{0;\text{imean}}$ and $V_{0;\text{i}}$ is shown in Fig. 2.17.

Comment

The term $R_i \cdot I_L$ in equation (2.11) represents a d.c. volt drop only, it does not have any associated power loss component.

When operating in inverting mode care must be taken to ensure that commutation is completed before the voltage crossover point. For this reason an extinction angle (ε) is sometimes specified such that:

$$\varepsilon = \beta - \gamma \tag{2.12}$$

2.3.4 *Half-controlled converters*

Figure 2.18(a) shows a three-phase, half-controlled full-bridge converter. This is similar to the three-phase, fully controlled bridge converter of Fig. 2.10(a) but with half the thyristors replaced by diodes. In addition, a freewheeling or commutating diode has been included in parallel with the load to prevent the reversal of load voltage and to assist in the commutation of the switching elements of the bridge. Figures 2.18(b) and 2.18(c) show the output voltage waveforms for the bridge assuming a constant current and instantaneous current transfer between devices at firing angles less than and greater than 60°. The output voltage under these circumstances is:

$$V_0 = \frac{3}{2\pi}\left(\int_{\alpha+\frac{\pi}{3}}^{\frac{2\pi}{3}} \hat{V}\sin\omega t\, d\omega t + \int_{\frac{\pi}{3}}^{\frac{2\pi}{3}+\alpha} \hat{V}\sin\omega t\, d\omega t\right) = \frac{3}{2\pi}\hat{V}(1+\cos\alpha) \quad \alpha < 60° \tag{2.13}$$

$$V_0 = \frac{3}{2\pi}\int_{\alpha}^{\pi} \hat{V}\sin\omega t\, d\omega t = \frac{3}{2\pi}\hat{V}(1+\cos\alpha) \quad \alpha > 60° \tag{2.14}$$

2.3.5 *Converter power factor*

Power factor is defined as:

$$\text{Power factor} = \frac{\frac{1}{T}\int_0^T v \cdot i\, dt}{V_{rms} \cdot I_{rms}} = \frac{\text{Mean power}}{V_{rms} \cdot I_{rms}} \tag{2.15}$$

A converter draws a non-sinusoidal current at supply frequency from the supply which can be represented by its fundamental component at supply frequency and a series of harmonics. Assuming that the a.c. system voltage remains sinusoidal then power will only be associated with fundamental frequency when:

$$\text{Power} = V_{1;rms} \cdot I_{1;rms} \cdot \cos\phi_1 \tag{2.16}$$

Substituting in equation (2.14) gives:

$$\text{Power factor} = \frac{V_{1;rms} \cdot I_{1;rms} \cdot \cos\phi_1}{V_{1;rms} \cdot I_{rms}} = \frac{I_{1;rms} \cdot \cos\phi_1}{I_{rms}} = \mu\cos\phi_1 \tag{2.17}$$

in which $\mu (= I_{1;rms}/I_{rms})$ is referred to as the **current distortion factor**.

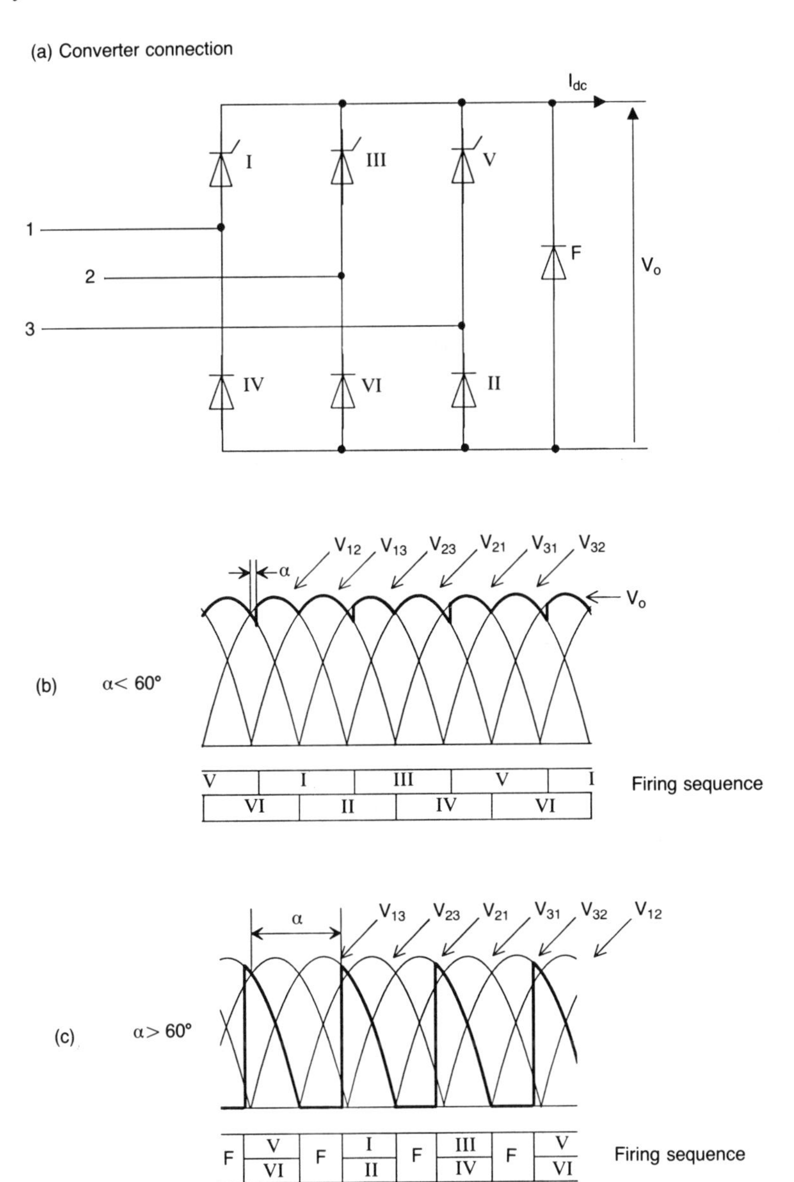

Fig. 2.18 Half-controlled, three-phase bridge converter: (a) converter connection; (b) $\alpha<60°$; and (c) $\alpha>60°$.

Whenever harmonic currents are present μ will be less than 1, even if the fundamental current and voltage are in phase ($\cos\phi_1=1$). For a fully controlled converter with a constant load current and ignoring overlap then $\cos\phi_1=\cos\alpha$. This means that a converter must be supplied with VAR by the supply system in order to operate.

Example 2.3 A d.c. link
Figure 2.19 illustrates the use of a d.c. link made up of two six-pulse, fully controlled bridge converters and a d.c. transmission system to connect two a.c. systems of different

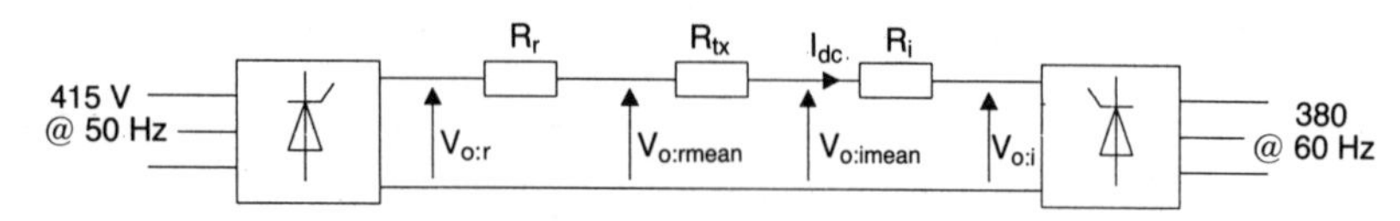

Fig. 2.19 Example 2.3: d.c. link.

voltages and frequency. Given that the source inductance of the 50 Hz system is 1 mH/phase and that of the 60 Hz system is 1.25 mH/phase find the firing angle of the rectifying converter and the firing advance angle of the inverting converter when the current in the d.c. system is 50 A and 15 kW is being delivered to the 60 Hz system.

Referring to equation (2.8) for the rectifier

$$\frac{p\omega L}{2\pi}=\frac{6\times 100\times \pi\times 10^{-3}}{2\pi}=0.3=R_{\mathrm{r}}$$

and equation (2.8) for the inverter

$$\frac{p\omega L}{2\pi}=\frac{6\times 120\times \pi\times 1.25\times 10^{-3}}{2\pi}=0.45=R_{\mathrm{i}}$$

Referring to the inverter, the mean voltage ($V_{\mathrm{i;mean}}$) can be found by reference to the power and the d.c. current when:

$$V_{\mathrm{i;mean}}=\frac{15\,000}{50}=300 \text{ volts}$$

The mean inverter voltage in the absence of overlap ($V_{\mathrm{i;0}}$) is then, using equation (2.11)

$$V_{\mathrm{i;0}}=V_{\mathrm{i;mean}}-R_{\mathrm{i}}\,.\,I_{\mathrm{L}}=300-50\times 0.45=277.5 \text{ volts}$$

From equation (2.9)

$$\cos\beta=\frac{275.5}{380\sqrt{2}}\times\frac{\pi}{6\sin 30^\circ}=0.5408$$

thus

$$\beta=48.53^\circ$$

Now

$$V_{\mathrm{r;mean}}=V_{\mathrm{i;mean}}+I_{\mathrm{L}}\,.\,R_{\mathrm{tx}}=300+50\times 0.2=310 \text{ volts}$$

From equation (2.9)

$$V_{\mathrm{r;0}}=V_{\mathrm{r;mean}}+R_{\mathrm{r}}\,.\,I_{\mathrm{L}}=310+50\times 0.3=325 \text{ volts}$$

From equation (2.6)

$$\cos\alpha=\frac{325}{415\sqrt{2}}\times\frac{\pi}{6\sin 30^\circ}=0.5799$$

thus

$$\alpha=54.56^\circ$$

2.4 Converters with discontinuous current

2.4.1 *Load inductance*

Where the load inductance is insufficient to maintain a continuous d.c. current, the output current will contain a ripple component which will be reflected in the supply current in the manner of Fig. 2.20. Analysis of both the converter and load conditions is now much more complex and requires the values of the system components to be taken into account.

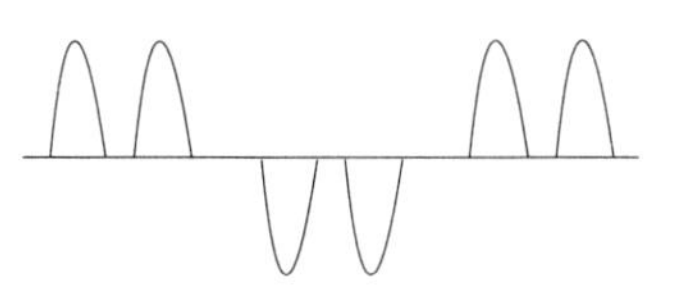

Fig. 2.20 Discontinuous phase current, six-pulse bridge.

2.4.2 *Capacitive smoothing*

Where capacitive smoothing is used on the output of a converter the resulting output voltage waveform will be of the general form shown in Fig. 2.21 for a two-phase uncontrolled bridge converter in which the diodes will start to conduct when the supply voltage exceeds the capacitor voltage. Conduction will then continue until the supply voltage falls below the capacitor voltage.

Comment

The above assumes ideal diodes. In practice, the forward volt drop of the diodes must be taken into account.

The current drawn from the supply will have the general form shown in Fig. 2.21(b) and will become increasingly 'peaky' as the load on the rectifier increases.

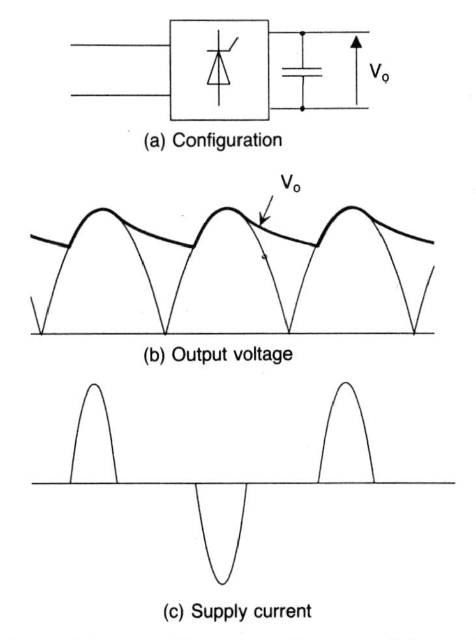

Fig. 2.21 Converter with capacitive smoothing: (a) configuration; (b) output voltage; and (c) supply current.

Comment

In a practical system differences between individual circuit components will result in asymmetry between the positive and negative current pulses and under very light loads may even result in current being drawn from the supply on only the positive or negative half cycle of voltage.

2.4.3 *Voltage bias*

Where a converter is used to supply a load such as a battery charger or a d.c. machine where the armature inductance is insufficient to ensure continuous current, the battery voltage or the back e.m.f. of the machine will appear as a bias voltage on the d.c. side of the converter. The operation of the converter under these circumstances then depends upon the relationship between the firing angle (α) of the thyristors, the point-on-wave at which the supply voltage exceeds the bias voltage (ψ) and the point-on-wave at which the thyristor stops conducting (σ). For the single-phase fully controlled bridge converter of Fig. 2.22 with $\alpha > \psi$, this will result in an output voltage waveform of the form shown for which:

$$V_0 = \frac{1}{\pi}\left(\int_{\alpha}^{\sigma} \hat{V} \sin \omega t \, d\omega t + \int_{\sigma-\pi}^{\alpha} V_{bias} \, d\omega t\right)$$

$$= \frac{1}{\pi}[\hat{V}(\cos \alpha - \cos \sigma) + V_{bias}(\alpha + \pi - \sigma)] \quad (2.18)$$

Similar solutions would exist for different combinations of firing angle and bias voltage.

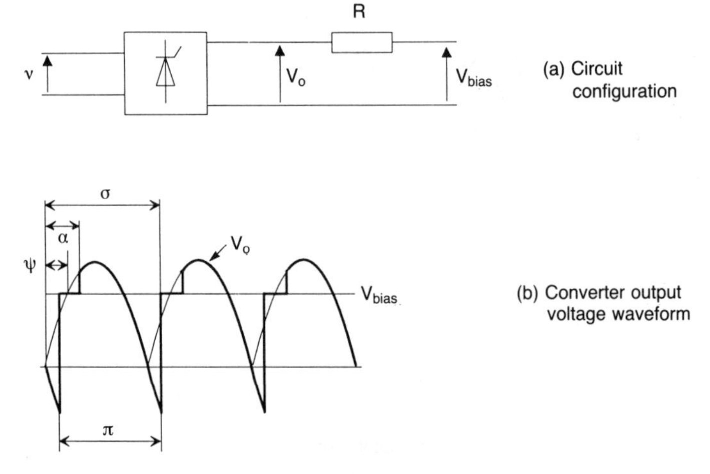

Fig. 2.22 Single phase bridge converter with voltage bias: (a) circuit configuration; and (b) converter output voltage waveform.

2.5 Choppers

A variable d.c. voltage can be obtained by rapidly turning on and off or **chopping** a constant d.c. scource voltage as in Fig. 2.23. The magnitude of the output voltage is determined by the mark/space ratio according to the relationship:

$$V_L = V_s\left(\frac{t}{T}\right) \tag{2.19}$$

For a chopper circuit such as that shown in Fig. 2.24(a), natural commutation no longer occurs and the switching device must be self-commutating such as the GTO thyristor shown. If a conventional thyristor is used then an external forced commutation circuit must be used in conjunction with the thyristor to turn it off.

The form of the load current is dependent upon the value of load inductance and hence the time constant of the load and the chopping frequency used. For the condition where the load time constant is large in relation to the chopping frequency the variation in load current may be considered to be linear and the conditions of Fig. 2.24(b) apply in which case, during conduction:

$$V_s - V_R = L\frac{di}{dt} \approx L\frac{\Delta i}{\Delta t} \tag{2.20}$$

when

$$I_1 - I_2 = \frac{(V_s - V_R).t_1}{L} \tag{2.21}$$

and during the off period $V=0$ and (2.22)

$$I_1 - I_2 = \frac{V_R(T-t_1)}{L} = \frac{t_2 . V_R}{L} \tag{2.23}$$

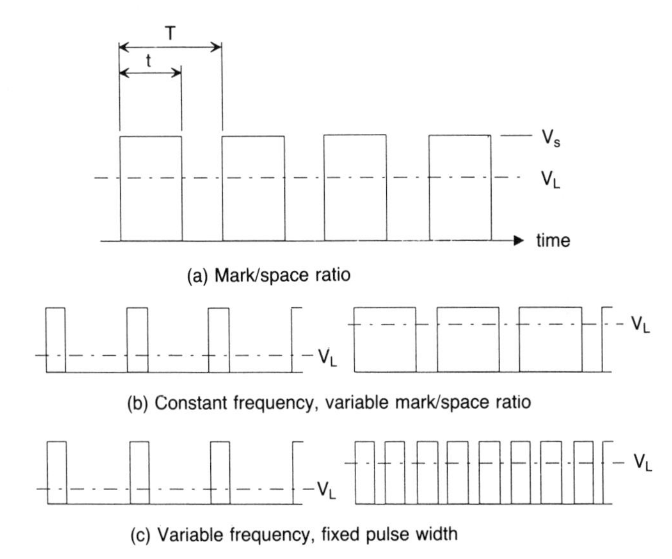

Fig. 2.23 Chopped voltage waveform: (a) mark/space ratio; (b) constant frequency, variable mark/space ratio; and (c) variable frequency, fixed pulse width.

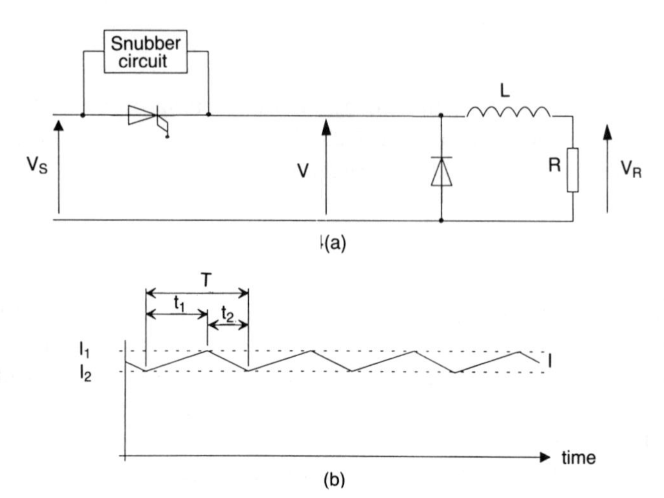

Fig. 2.24 (a) Circuit configuration and (b) chopper output current, system time constant much greater than T.

Also

$$I_{\text{mean}} = \frac{I_1 + I_2}{2} \tag{2.24}$$

The ripple current (i_r) is then:

$$i_r = (I_1 - I_2).\left(\frac{t}{t_1} - \frac{1}{2}\right), \qquad \text{for } 0 < t < t_1 \tag{2.25}$$

and

$$i_r = (I_1 - I_2).\left(\frac{1}{2} - \frac{(t - t_1)}{t_2}\right), \qquad \text{for } t_1 < t < T \tag{2.26}$$

When operating under these conditions care must be taken to ensure that the di/dt limit of the switching device is not exceeded, particularly on turn-on when current is transferred from the freewheeling diode. To limit the effects the snubber circuit should be designed accordingly, including the possibility of introducing series inductance as in Fig. 2.25.

Referring to Fig. 2.24, for the case where the period T is of the same order as the time constant of the load the variation of current can no longer be considered to be linear, in which case, during conduction:

$$i_L = I_2 + \left(\frac{V_s}{R} - I_2\right)(1 - e^{-Rt/L}) \tag{2.27}$$

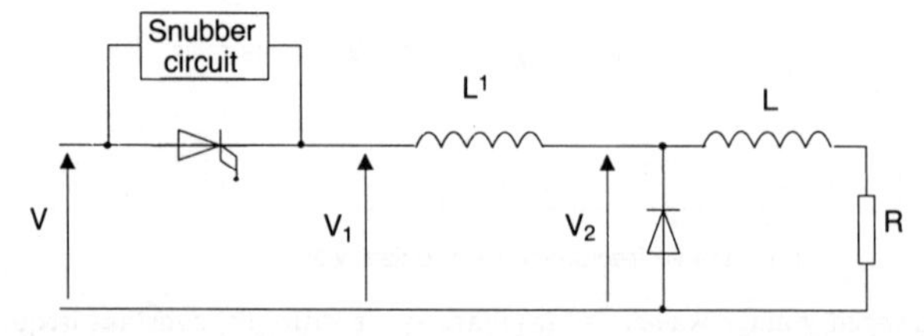

Fig. 2.25 Chopper with series inductance in source.

and

$$I_1 = I_2 + \left(\frac{V_s}{R} - I_2\right)(1 - e^{-Rt_1/L}) \tag{2.28}$$

and, during the off period

$$i_L = I_1 \,.\, e^{-Rt/L} \tag{2.29}$$

and

$$I_2 = I_1 \,.\, e^{-Rt_2/L} \tag{2.30}$$

Comment

In equations (2.26) and (2.28) time t is measured from the appropriate transition point on the waveform.

The circuits of Figs 2.24 and 2.25 only permit power to flow from the supply to the load. Such a circuit is referred to as a class A or single-quadrant chopper and operates in the first quadrant of the v_L/i_L diagram of Fig. 2.26.

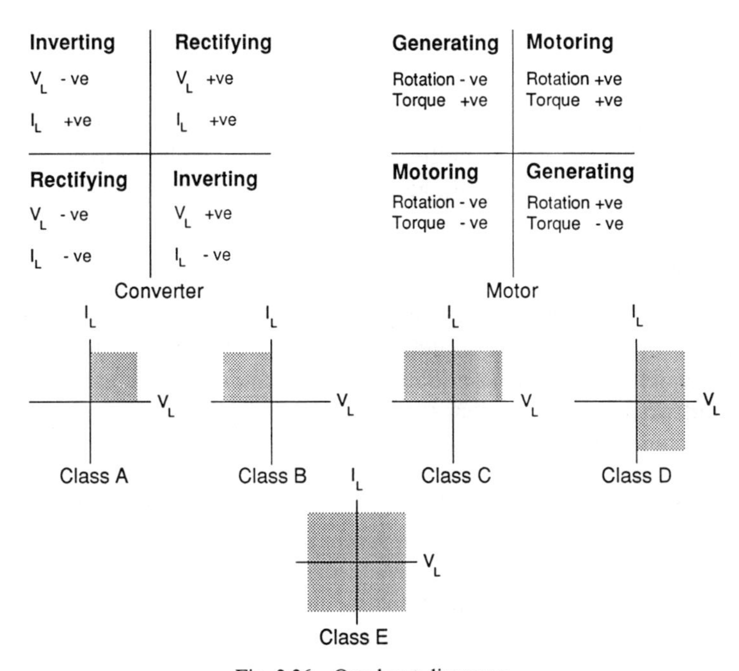

Fig. 2.26 Quadrant diagrams.

Example 2.4 d.c. chopper

A class A chopper of the type shown in Fig. 2.24(a) is operating at a frequency of 2 kHz from a 120 volt d.c. source to supply a load with a resistance of 8 ohms. The load time constant (τ) is 6 ms. If the mean load voltage is 72 volts, find the required mark/space ratio for the chopper, the mean load current and the magnitude of the current ripple.

$$\text{Period} = T = \frac{1}{f} = \frac{1}{2000} = 0.5 \text{ ms}$$

As the load time constant is $>10T$ the current variation can be considered as linear when, from equation (2.18):

$$t_1 = T.\frac{V_R}{V_s} = \frac{0.5 \times 10^{-3} \times 72}{120} = 0.3 \text{ ms}$$

The chopped waveform therefore has a mark/space ratio of 60%. The mean load current is then

$$I_{mean} = \frac{V_R}{R} = \frac{72}{8} = 9 \text{ A}$$

The load inductance is obtained from the time constant as $L = \tau R = 6 \times 10^{-3} \times 8 =$ 48 mH. From equation (2.22)

$$I_1 - I_2 = \frac{t_2 . V_R}{L} = \frac{0.2 \times 10^{-3} \times 72}{48 \times 10^{-3}} = 0.3 \text{ A}$$

2.6 Inverters

An inverter is used to produce a variable frequency a.c. supply from a d.c. source by means of the controlled switching of the source using either self or forced commutated devices. Inverters may be either voltage sourced or current sourced.

2.6.1 *Voltage sourced inverters*

Figure 2.27 shows the basic circuit configuration for a single-phase voltage sourced inverter using transistors as the switching elements. The reverse diodes across the transistors are included to accommodate the out of phase current associated with an inductive load when the transistors (or other switching device) are turned off. The capacitor is included at the input for voltage regulation and to absorb any returned energy from an inductive load.

If the pairs of transistors T_1/T_2 and T_3/T_4 are turned on together and in sequence an alternating voltage is produced across the load, the exact form of which depends upon the conduction period of the transistors as shown in Fig. 2.28 together with the associated current waveforms for an inductive load.

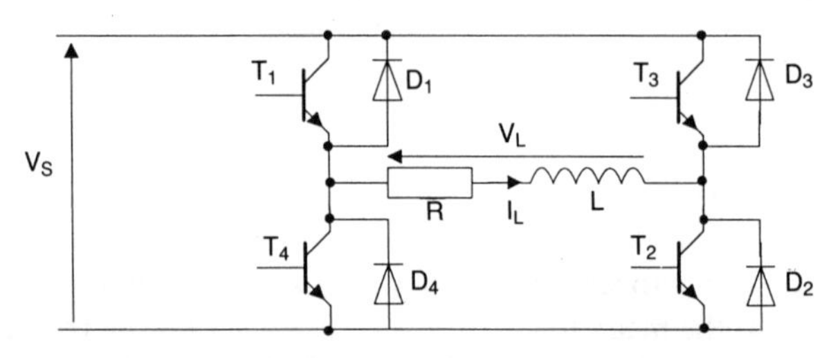

Fig. 2.27 Single-phase voltage sourced inverter.

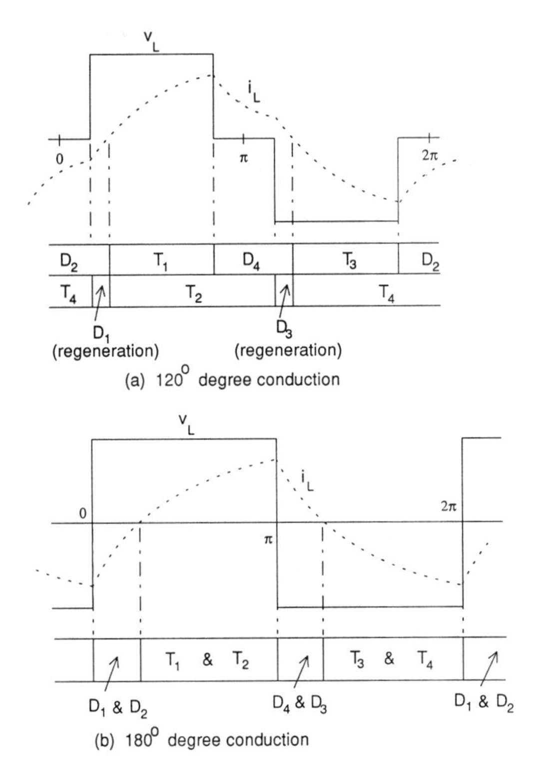

Fig. 2.28 Output current and voltage waveforms for a single-phase voltage sourced inverter: (a) 120° conduction; and (b) 180° conduction.

Comment

Care must be taken in determining the control strategy for the inverter to ensure that pairs of devices such as T_1/T_4 and T_3/T_2 which would form a short circuit across the supply are never conducting simultaneously.

Figure 2.29 similarly shows a three-phase voltage sourced transistor inverter together with the output voltage waveforms for a star connected load with device conduction periods of 120° and 180°.

Example 2.5 Three-phase bridge inverter
A three-phase bridge inverter such as that shown in Fig. 2.29 is connected to a 660 volt d.c. source and is used to supply a balanced, three-phase star connected load of 20 ohms/phase. Find the r.m.s. load current and the load power for (a) 120° conduction and (b) 180° conduction.

(a) For 120° conduction:

$$\text{Load current} = I_L = \frac{660}{2 \times 20} = 16.5 \text{ A}$$

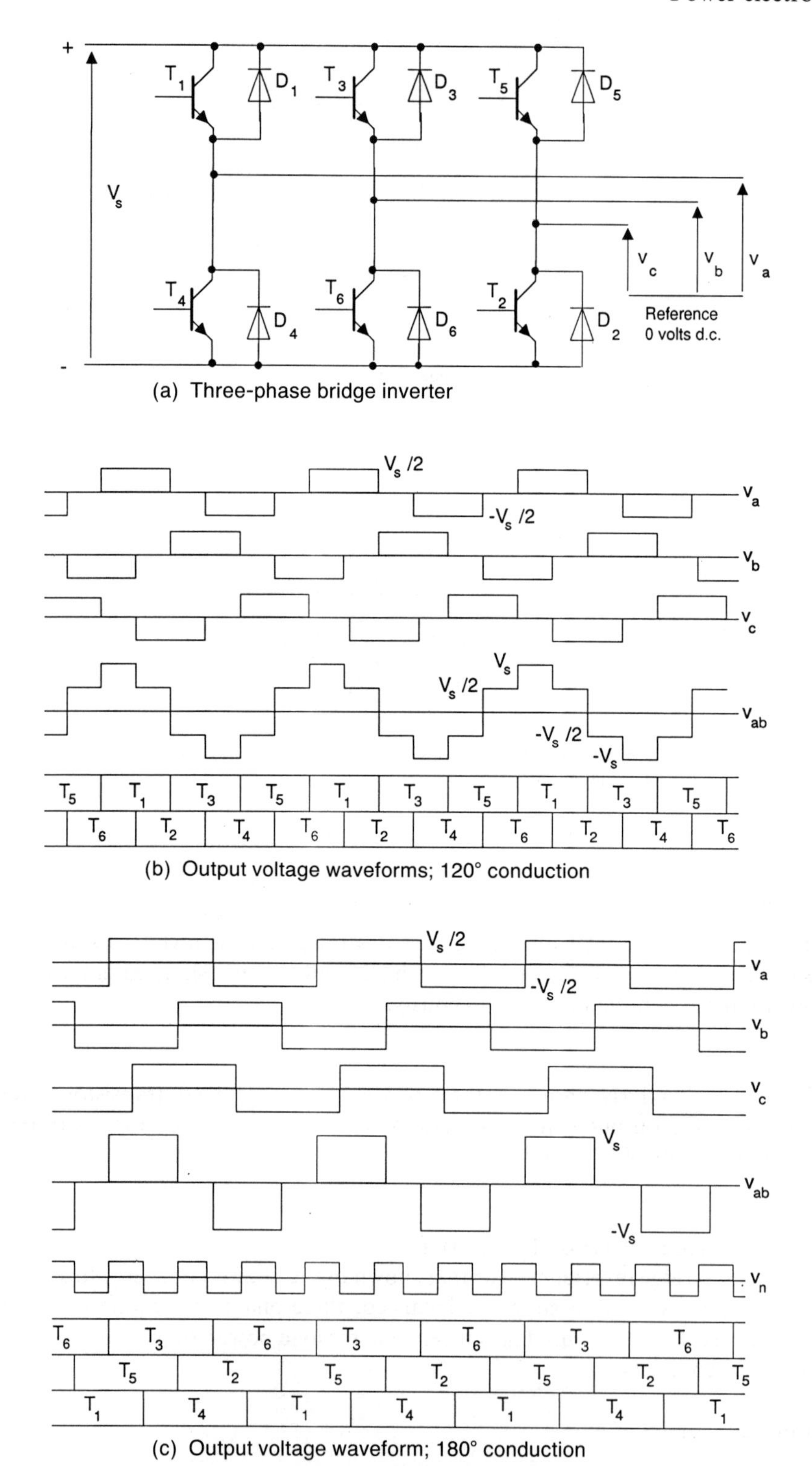

(a) Three-phase bridge inverter

(b) Output voltage waveforms; 120° conduction

(c) Output voltage waveform; 180° conduction

Fig. 2.29 (a) Three-phase voltage sourced bridge inverter, (b) output voltage waveforms (120° conduction) and (c) output voltage waveform (180° conduction).

and

$$I_{\mathrm{Lrms}}=\left[\frac{1}{2\pi}\left(\int_0^{\frac{2\pi}{3}} I_L^2\,d\theta+\int_{\pi}^{\frac{5\pi}{6}} I_L^2\,d\theta\right)\right]^{\frac{1}{2}}=\left[\frac{(16.5^2+16.5^2)}{3}\right]^{\frac{1}{2}}=13.47\ \mathrm{A}$$

Load power $=13.47^2\times 20\times 3=10\,890$ watts

Thyristor r.m.s. current $=I_{\mathrm{Trms}}=\left(\frac{16.5^2}{3}\right)^{\frac{1}{2}}=9.52$ A

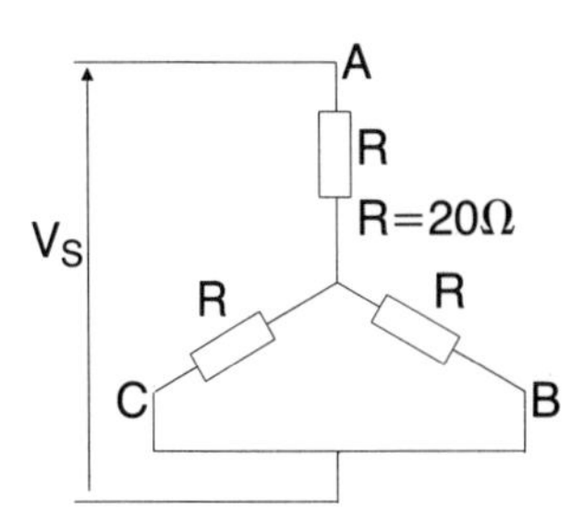

Fig. 2.30 Effective connection of load for 180° conduction.

(b) For 180° conduction, at any instant the load is effectively connected in the manner shown in Fig. 2.30. The effective load on the inverter is then:

$$R_{\mathrm{load}}=20+\frac{20}{2}=30\ \text{ohms}$$

and

$$I_1=\frac{660}{30}=22\ \mathrm{A}$$

and

$$I_2=\frac{I_1}{2}=11\ \mathrm{A}$$

$$I_{\mathrm{Lrms}}=\left[\frac{1}{2\pi}\left(\int_0^{\frac{2\pi}{3}} I_2^2\,d\theta+\int_{\frac{2\pi}{3}}^{\frac{4\pi}{3}} I_1^2\,d\theta+\int_{\frac{4\pi}{3}}^{\beta\pi} I_2^2\,d\theta\right)\right]^{\frac{1}{2}}=\left(\frac{2\times 11^2+22^2}{3}\right)^{\frac{1}{2}}=15.56\ \mathrm{A}$$

Load power $=15.56^2\times 20\times 3=14\,530$ watts

As the thyristors carry current I_1 for one-sixth of a cycle and current I_2 for one third of a cycle the r.m.s. current in the thyristors is:

$$I_{\mathrm{Trms}}=\frac{I_{\mathrm{Lrms}}}{\sqrt{2}}=11\ \mathrm{A}$$

2.6.2 *Harmonics*

The output voltage of inverters such as those shown in Figs 2.27 and 2.29 is in the form of a square or quasi-square wave with a high harmonic content, the presence of which results in increased I^2R losses as a result of the increase in r.m.s. current, increased losses in magnetic materials and harmonic torque generation in machines. The use of the pulse-width modulated (PWM) inverter affords a means of reducing the harmonic content of the inverter output voltage waveform, and hence of the output current, thus improving overall system performance.

2.6.3 *Pulse width modulated inverters*

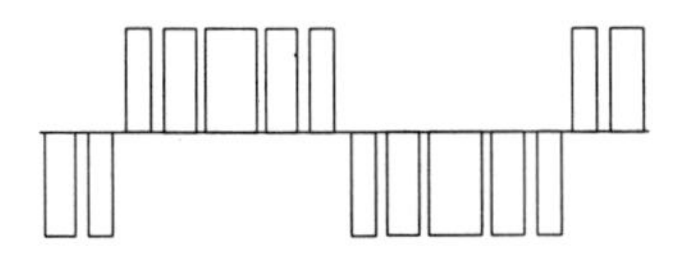

Fig. 2.31 General pulse width modulated waveform.

Applications such as induction machines require control of both the magnitude and the frequency of the a.c. voltage produced by the inverter. By using a pulse width modulated pulse train of the general form shown in Fig. 2.31 in which the $\int v\,.dt$ relationship approximates to that of a sinewave, a near sinusoidal current can be produced in the load. Figure 2.32 demonstrates the generation of a pulse width modulated waveform by means of a reference triangular waveform, while Fig. 2.33 shows the form of the associated current waveform for an inductive load.

The form of the PWM waveform may be defined in terms of the **modulation index** (δ_m), defined as the ratio of the actual magnitude of the fundamental component to the maximum value of fundamental that could be obtained, and the **frequency ratio** or **gear ratio**, defined as the number of pulses per cycle of the fundamental of the output waveform.

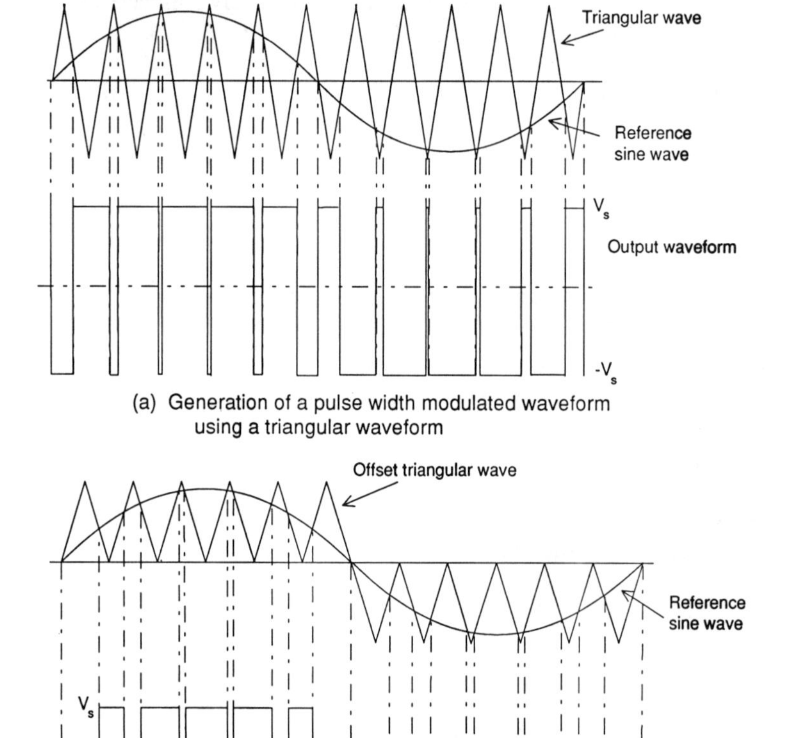

Fig. 2.32 Production of pulse width modulated waveform: (a) using a triangular waveform; and (b) using an offset triangular waveform.

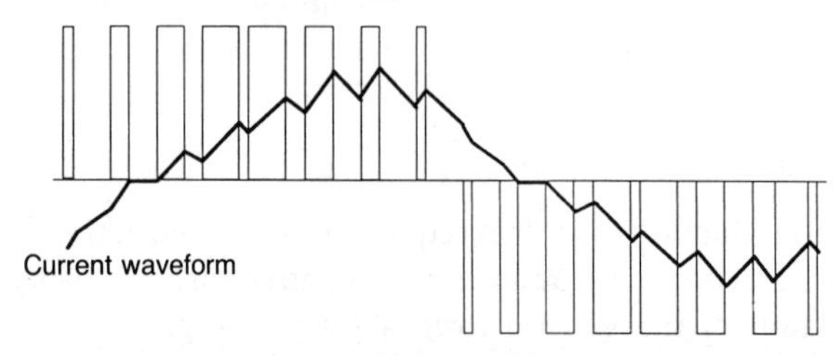

Fig. 2.33 Current waveform from a pulse width modulated inverter with an inductive load.

(a) *Harmonic elimination*

The shape of the PWM waveform may be controlled to eliminate specific harmonics in the voltage waveform in which case m chops per quarter cycle can be used to eliminate m harmonics in the output voltage waveform or, alternatively, to control the modulation index together with $m-1$ harmonics. Implementation is typically by means of tables of precalculated values of switching angles covering a range of frequencies and stored in the form of look-up tables for fast access. However, as computation speeds increase for the same or reducing cost it may become possible to calculate the required firing angles in real time according to an appropriate strategy for harmonic elimination and control, further increasing inverter performance.

2.6.4 *Current-sourced inverter*

In a current-sourced inverter, the current drawn from the d.c. source is maintained at an effectively constant level by means of a large indutance connected in series with the d.c. supply which enables changes of inverter voltage to be accommodated at low values of di/dt. Current-sourced inverters are typically used to supply high power factor loads whose impedance either remains constant or decreases at harmonic frequencies, thus preventing possible problems either on switching or with harmonic overvoltages. A particularly useful feature of the current-sourced inverter in respect of machine control is the ability to directly control current and hence torque by using a controlled converter as the d.c. source.

The principle of operation of a current-sourced inverter using thyristors is illustrated by Fig. 2.34 and shows the simple combination of diodes and capacitors necessary for forced commutation. With thyristors T_1 and T_2 conducting, the capacitors C_1 and C_2 will be charged as shown. When thyristors T_3 and T_4 are fired, the capacitors will discharge through the loops formed by T_1 and T_3, and T_1 and T_4 respectively, turning T_1 and T_2 off. Once T_1 and T_2 are turned off, the capacitors will then continue to discharge and then charge in the reverse direction via the path T_3–C_1–D_1–load–D_2–C_2–T_4. When the capacitors are fully charged in the reverse direction diodes D_3 and D_4 will start to conduct, reversing the direction of load current and completing the transfer of load to thyristors T_3 and T_4.

The principle of the single-phase current-sourced inverter can readily be extended to a three-phase configuration as illustrated by Fig. 2.35 in which case the thyristors fire in the sequence T_1–T_2–T_3–T_4–T_5–T_6–T_1, etc., with each thyristor conducting for 120°.

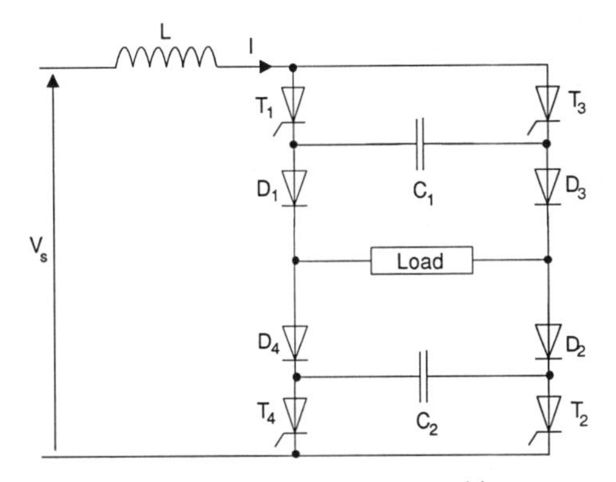

Fig. 2.34 Single-phase current sourced inverter.

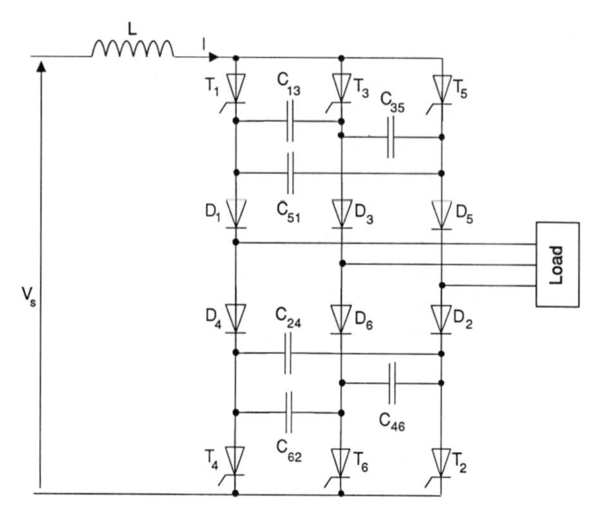

Fig. 2.35 Three-phase current sourced inverter.

Comment

PWM techniques are not generally used with current-sourced inverters except for possible applications in the control of machines at low speeds where there is some indication that its use improves stability.

(a) *Inverter performance*

The major source of loss in an inverter is likely to be the switching loss incurred during the transition from the on-state to the off-state of each switching device and *vice-versa*. However, in addition to the switching losses there will be losses in the snubber circuit and, in the case of a force-commutated (thyristor) inverter there will be additional losses in the commutation circuit.

Typical efficiencies for voltage-sourced inverters are of the order of 96% for a quasi-square wave inverter, including the converter and d.c. link, 94% for a PWM inverter using self-commutated devices and 91% for a PWM inverter using thyristors with forced commutation.

Output frequencies range from a few hertz to around 100 Hz for quasi-square wave inverters using thyristors, though this latter figure may be higher for special applications, to 500 Hz if self-commutating devices are used. The maximum output frequency of PWM inverters is limited by switching losses in the devices to frequencies around 100 Hz.

Current sourced inverters typically operate in a frequency range from a few hertz to the supply frequency, although supersynchronous operation at higher frequencies is possible. The upper limit for frequency is set by the time required for commutation. Commutation losses are small and overall efficiencies are of the order of 96%.

2.7 Cycloconverters

A cycloconverter synthesizes its output voltage waveform by switching between the phases of the a.c. supply and provides an alternative to an inverter where the required output frequencies are restricted to less than the supply frequency. Figure 2.36 shows the basic configuration of a single-phase cycloconverter.

Cycloconverters require more complex control systems than inverters which tends to limit their application to high power ratings. They are, however, very flexible in operation and four-quadrant operation is readily achieved.

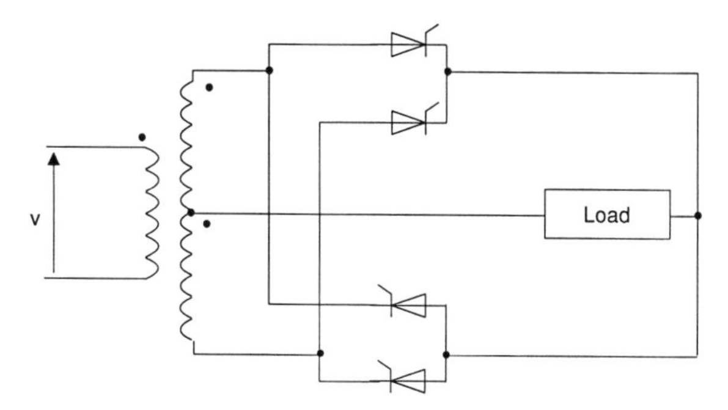

Fig. 2.36 Single-phase cycloconverter.

2.8 Summary

Power electronic switching devices have an important role to play in the operation and control of electric machines and their application with specific types of machine is considered in the appropriate chapters. Figure 2.37 provides a summary of the various circuit configurations considered in this chapter and gives the appropriate schematic form of circuit representation that will be used in the remainder of the text.

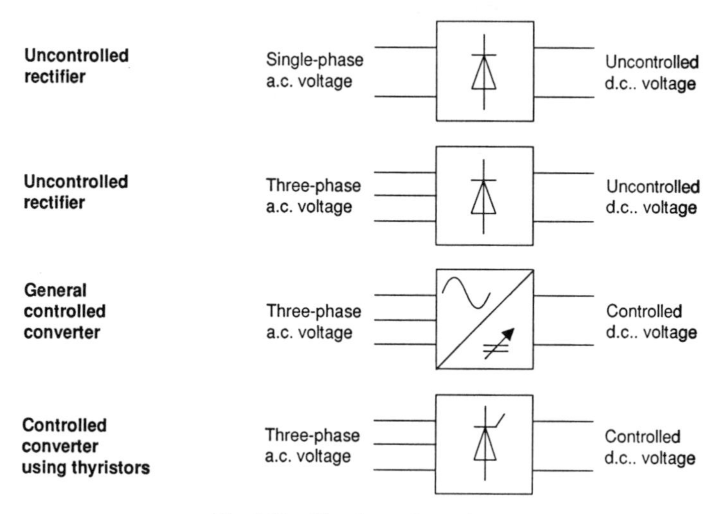

Fig. 2.37 Circuit configurations.

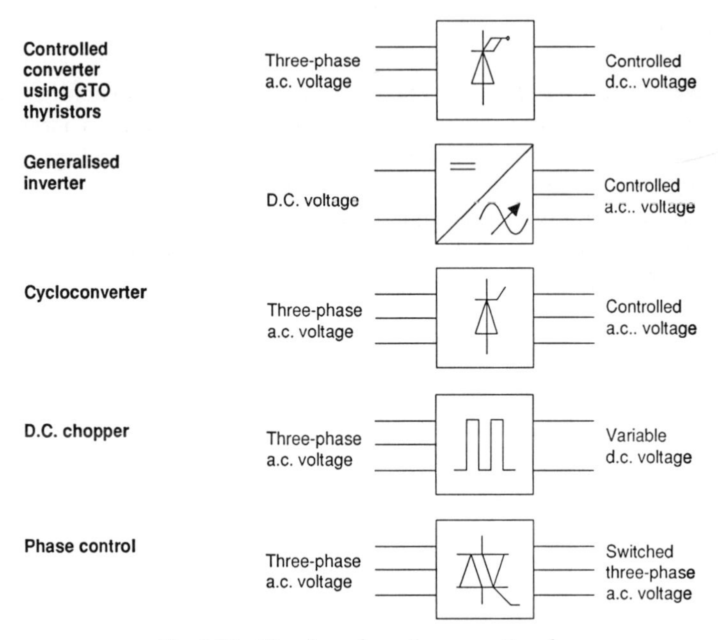

Fig. 2.37 Circuit configurations—*continued.*

Exercises

1 The gate drive circuit and gate characteristics for a particular thyristor are shown in Fig. 2.38. Obtain a value for the resistance R that will ensure firing of the thyristor when the transistor is turned on.

2 The full-wave, fully controlled, single-phase bridge converter of Fig. 2.39 is used to supply a load of resistance 3.6 Ω and inductance 480 mH from a 240 volt (r.m.s.), 50 Hz a.c. supply. For a firing angle of 30° find (a) the average and r.m.s. currents in the load and (b) the average and r.m.s. currents in the thyristors. The thyristor forward volt drops may be neglected.

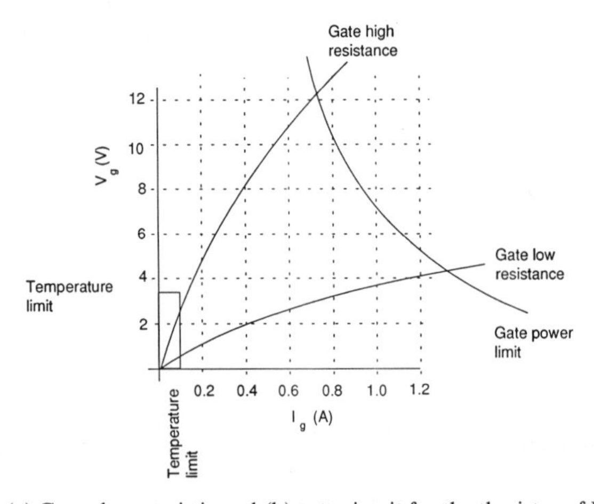

Fig. 2.38 (a) Gate characteristic and (b) gate circuit for the thyristor of Exercise 1.

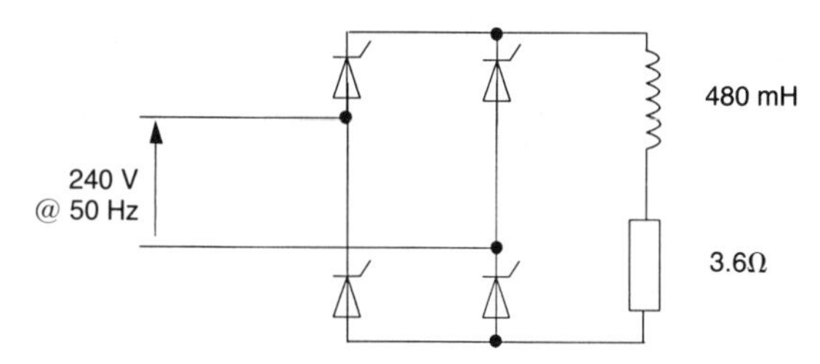

Fig. 2.39 Single-phase fully controlled bridge converter.

3 If a free-wheeling diode is now fitted across the load in Fig. 2.39 show that the output voltage becomes:

$$V_0 = 0.3183\ \hat{V}_s(1+\cos\alpha) = 108(1+\cos\alpha)$$

For the same load and firing angle as in Exercise 2 find (a) the average and r.m.s. currents in the load and (b) the average and r.m.s. currents in the thyristors.

4 A six-pulse, fully controlled bridge converter is connected to a three-phase, 50 Hz, 415 volt (line) a.c. supply and to a d.c. generator. Find the firing advance angle (β) and the mean and r.m.s. thyristor currents when the generator is providing 30 A (d.c.) and 250 volts (d.c.). The generator has a resistance of 1.2 Ω and sufficient inductance to ensure a continuous, steady d.c. current.

5 A six-pulse, fully controlled bridge converter is connected to a three-phase, 50 Hz, 415 volt (line) a.c. supply. If thyristors with the characteristics given are used in the bridge, what will be the maximum power rating of the bridge? If each thyristor is mounted on a separate heat sink, determine the thermal resistance of the heat sink assuming a maximum value for the ambient temperature of 30 °C.

Forward volt-drop in on state	= 1.28 V
Average on state current	= 25 A maximum
R.M.S. on state current	= 42 A maximum
Continuous on state current	= 36 A maximum
Peak duty cycle current	= 250 A
Maximum junction temperature	= 130 °C
Junction to heat sink thermal resistance	= 1.16 °C/W

What will be the power factor associated with the operation of a fully controlled, single-phase bridge converter at firing angles of (a) 30° and (b) 60°? Overlap and device volt-drops can be ignored.

A simple d.c. chopper is operated at a frequency of 2 kHz from a 96 volt d.c. source to supply a resistance of resistance 8 Ω and inductance 48 mH. If the mean load voltage is 57.6 volts, find the on time of the thyristor, the mean load current and the magnitude and r.m.s. values of the ripple current.

A three-phase, voltage sourced inverter is used to supply a star connected resistive load of 15 Ω/phase from a 600 volt d.c. source. Determine the r.m.s. load current, the load power and the thyristor r.m.s. current ratings for (a) 120° conduction and (b) 180° conduction.

3 Basic electromagnetism for electric machines and transformers

3.1 Ampère's law and Faraday's law

Ampère's law is expressed by equation (3.1) and relates the line integral of the **magnetic field intensity** (H A m^{-1}) around a closed contour to the current enclosed by that contour according to the relationship.

$$I = \oint \boldsymbol{H} . \mathrm{d}\boldsymbol{l} \tag{3.1}$$

The $\oint \boldsymbol{H} . \mathrm{d}\boldsymbol{l}$ term in equation (3.1) is referred to as the **magnetomotive force** or **m.m.f.** (F) and is expressed in units of ampere-turns.

Faraday's law is expressed by equation (3.2) and relates to the generation of an **electromotive force** or e.m.f. in a conductor in the presence of a time varying magnetic field.

$$e = -\frac{\mathrm{d}\Phi}{\mathrm{d}t} \tag{3.2}$$

Consider Fig. 3.1(a) in which a conductor length $\boldsymbol{l}$ is moving with a velocity $\boldsymbol{v}$ in a plane normal to a constant magnetic field with a **flux density** of $\boldsymbol{B}$ tesla. The area swept by the conductor in unit time is given by:

$$\text{Area/unit time} = A_t = \boldsymbol{l} . \boldsymbol{v} \tag{3.3}$$

The e.m.f. e induced in the conductor is, since $\Phi = \boldsymbol{B} . A$:

$$e = \boldsymbol{B} . \boldsymbol{l} . \boldsymbol{v} \tag{3.4}$$

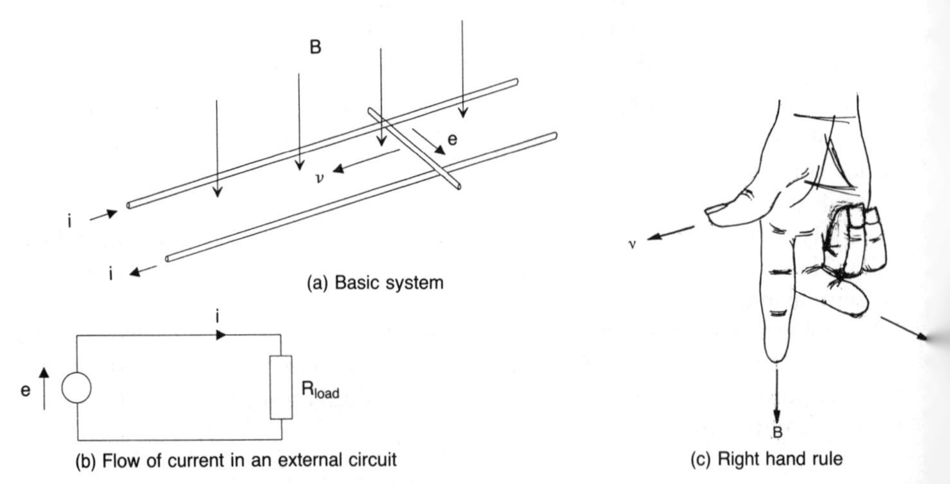

Fig. 3.1 Production of an e.m.f. in a conductor moving in a magnetic field (a) basic system; (b) flow current in an external circuit; and (c) right-hand rule.

in the direction shown. This would result in the flow of a current in an external circuit in the direction indicated in Fig. 3.1(b).

Comment

The sense of the induced voltage is determined by the **right-hand rule** shown in Fig. 3.1(c). The first finger points in the direction of the field, the thumb in the direction of motion of the conductor relative to the field and the second finger then indicates the sense of the induced voltage.

3.2 Force on a current carrying conductor in magnetic field

The force $\boldsymbol{F}$ on a current carrying conductor in a magnetic field is defined by the vector equation

$$\boldsymbol{F} = \boldsymbol{I} \times \boldsymbol{B} \tag{3.5}$$

The direction of this force may be determined by the use of the **left-hand rule** in which the first finger of the left hand is aligned with the direction of the magnetic field and the second finger, bent at right angles to the first as in Fig. 3.2(a), is aligned with the direction of the current. The thumb then points in the direction of the force. Referring to Fig. 3.2(b), the direction of the force on the conductor would then be as shown.

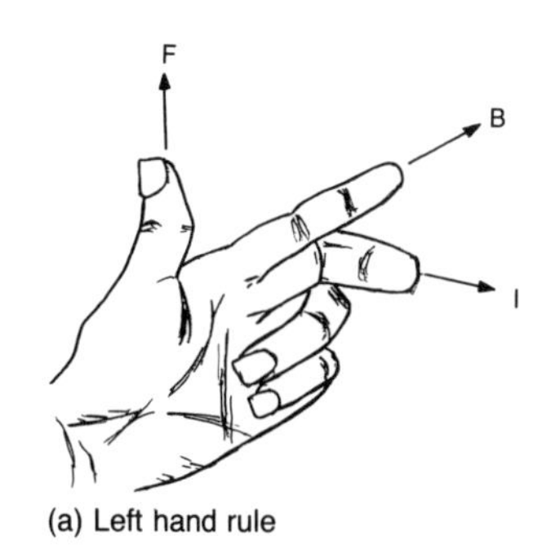

(a) Left hand rule

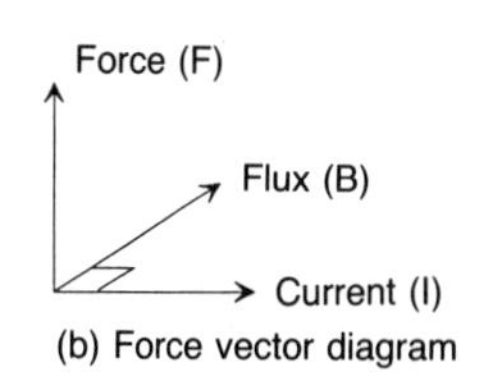

(b) Force vector diagram

Fig. 3.2 Force on a current-carrying conductor in a magnetic field: (a) left-hand rule; and (b) force vector diagram.

Comment

Referring back to Fig. 3.1 and applying the left-hand rule with current flowing in the external circuit, it will be seen that the force on the conductor that results opposes the motion that is generating the e.m.f. This is the set of conditions that exist in a **generator** where mechanical energy is being converted into electrical energy.

In the case of a motor, a current is driven through the conductor by an external applied voltage to produce a torque. The e.m.f. in the conductor that results from its motion through the field – often referred to as the **back e.m.f.** – then opposes the applied voltage to limit the conductor current.

3.3 Magnetic circuits

Consider the simple magnetic circuit of Fig. 3.3. The presence of a current I in the coil of N turns results in a magnetic flux $\boldsymbol{\Phi}$ in the material of the circuit such that:

$$F = N \cdot I = H_{mm} \cdot l_{mm} + H_{gap} \cdot l_{gap} \tag{3.6}$$

where $F = N \cdot I$ is the m.m.f. and l_{mm} and l_{gap} are the lengths of the flux path in the **magnetic material** and in the **air gap** respectively.

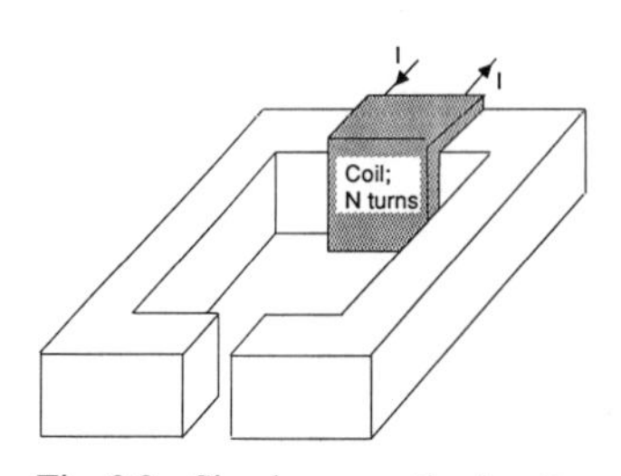

Fig. 3.3 Simple magnetic circuit.

In the magnetic material and assuming a straight line relationship between the magnetic field intensity ($\boldsymbol{H}_{mm}$) and the magnetic flux density ($\boldsymbol{B}$).

$$H_{mm} = \frac{B_{mm}}{\mu_o \cdot \mu_r} = \frac{\Phi}{A_{mm} \cdot \mu_o \cdot \mu_r} \tag{3.7}$$

where μ_r is the **relative permeability** of the magnetic material.
In the air gap

$$H_{gap} = \frac{B_{gap}}{\mu_o} = \frac{\Phi}{A_{gap} \cdot \mu_o} \tag{3.8}$$

Hence

$$F = \frac{\Phi . l_{mm}}{A_{mm} \cdot \mu_o \cdot \mu_r} + \frac{\Phi . l_{gap}}{A_{gap} \cdot \mu_o} = \Phi . (R_{mm} + R_{gap}) \tag{3.9}$$

where R_{mm} and R_{gap} are the **reluctances** of the flux path in the magnetic material and the air gap respectively.

Reluctance in a magnetic circuit is analogous to resistance in an electric circuit and the rules for the series and parallel combination of reluctances are the same as those for series and parallel combinations of resistances.

Example 3.1 The magnetic circuit
A magnetic circuit consists of a toroid of mean radius 15 cm with a 0.8 cm airgap. The cross-sectional area of the toroid is 1.2 cm^2 and it is made of a material with a relative permeability (μ_r) of 800. What current would be required in a coil of 720 turns wound on the toroid in order to produce a magnetic flux density of 0.72 tesla in the airgap?

Assume that the effective cross-sectional area of the airgap is 10% greater than that of the toroid to allow for fringing effects.

$$\Phi = B_{gap} \cdot A_{gap} = 0.72 \times 1.1 \times 1.2 \times 10^{-4} = 9.504 \times 10^{-5} \text{ Wb}$$

The flux density in the magnetic material is then

$$B_{mm} = \frac{\Phi}{A_{mm}} = \frac{9.504 \times 10^{-5}}{1.2 \times 10^{-4}} = 0.792 \text{ T}$$

Then

$$NI = H_{mm} \cdot l_{mm} + H_{gap} \cdot l_{gap} = \frac{B_{mm} \cdot l_{mm}}{\mu_o \mu_r} + \frac{B_{gap} \cdot l_{gap}}{\mu_o}$$

$$= \frac{0.792 \times 0.9425}{4\pi \times 10^{-7} \times 800} + \frac{0.72 \times 0.008}{4\pi \times 10^{-7}} = 5326.2$$

Thus

$$I = \frac{5326.2}{N} = 7.4 \text{ A}$$

3.3.1 *Energy stored in a magnetic field*

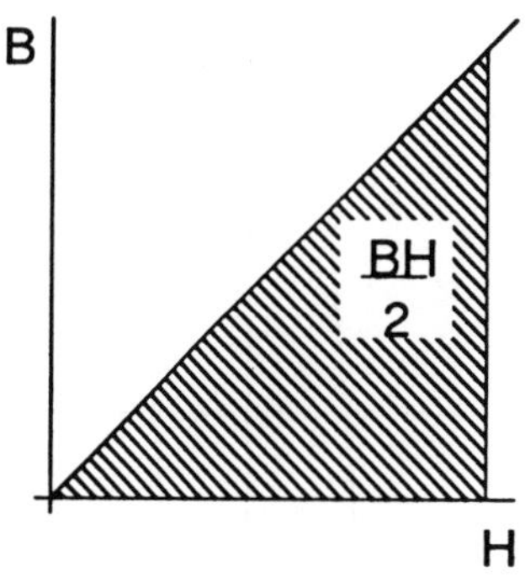

Fig. 3.4 Energy stored in a magnetic field.

Assuming a linear relationship between $\boldsymbol{B}$ and $\boldsymbol{H}$, the energy stored in a magnetic field of strength $\boldsymbol{B}$ associated with a field of $\boldsymbol{H}$ is then the area under the curve in Fig. 3.4 and is expressed by equation (3.10):

$$\text{Stored energy per unit volume} = \frac{1}{2}\,\boldsymbol{B}.\boldsymbol{H} \tag{3.10}$$

Comment

The total energy stored in a magnetic field can be expressed in a variety of forms using the relationships:

$$L = \frac{N\Phi}{I}$$

$$R = \frac{\boldsymbol{l}}{A\,.\,\mu_o\,.\,\mu_r}$$

$$\boldsymbol{B} = \frac{\boldsymbol{\Phi}}{A}$$

$$\boldsymbol{H}.\boldsymbol{l} = NI$$

where:

L is the inductance
R is the reluctance
A is the cross-sectional area of the magnetic circuit
$\boldsymbol{\Phi}$ is the magnetic flux
$\boldsymbol{l}$ is the path length

when

$$\text{Total stored energy} = \frac{1}{2}\,\boldsymbol{B}.H.A.\boldsymbol{l} = \frac{1}{2}\,LI^2 = \frac{1}{2}\,R\boldsymbol{\Phi}^2, \text{ etc.}$$

3.3.2 *Force between magnetic poles*

Consider the situation of Fig. 3.5. For a small movement of δx in the direction shown the work done against the force of attraction between the poles is equal to the change in stored energy that results. Assuming constant flux throughout then:

$$\boldsymbol{F}.\delta x = \frac{1}{2}\,\boldsymbol{\Phi}^2\,\delta R \tag{3.11}$$

when

$$\boldsymbol{F} = \frac{1}{2}\,\boldsymbol{\Phi}^2\,\frac{\delta R}{\delta x} = \frac{1}{2}\,\boldsymbol{\Phi}^2\,\frac{\mathrm{d}R}{\mathrm{d}x} \tag{3.12}$$

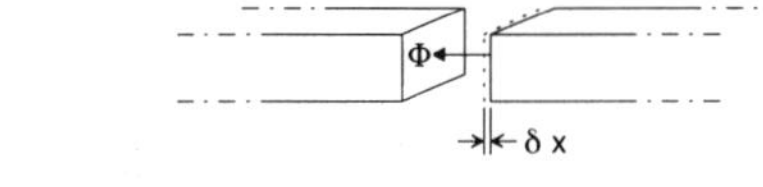

Fig. 3.5 Force of attraction between magnetic poles.

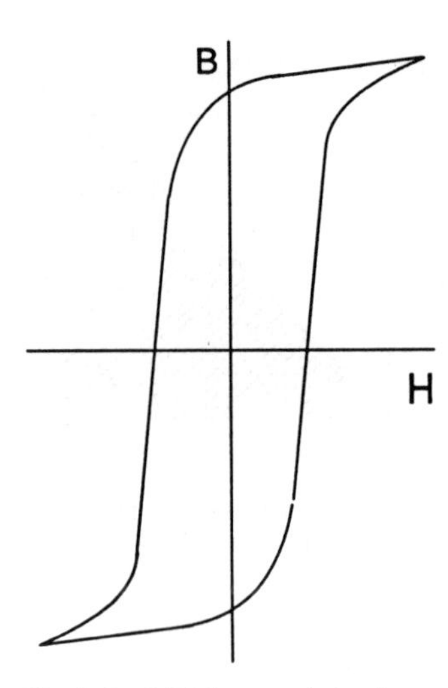

Fig. 3.6 *B*/*H* characteristic of a ferromagnetic material.

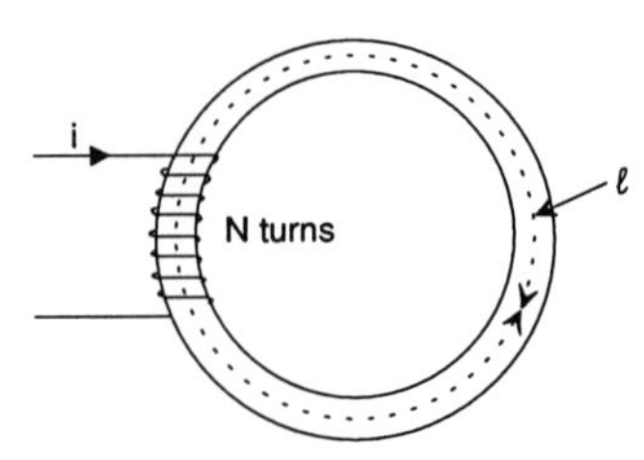

Fig. 3.7 Annular magnetic circuit.

Comment

Equation (3.12) can also be expressed in terms of the energy stored per unit volume $\left(\frac{1}{2}.\boldsymbol{BH}\right)$ and the cross-sectional area of the airgap (A) as:

$$\boldsymbol{F}=\frac{1}{2}.\boldsymbol{B}.\boldsymbol{H}.A$$

3.3.3 *The B/H loop*

The analysis of the magnetic circuit in the previous section assumed a constant permeability or linear relationship between $\boldsymbol{B}$ and $\boldsymbol{H}$ in the magnetic material. In practice, the relationship between $\boldsymbol{B}$ and $\boldsymbol{H}$ is complex and is expressed by the ***B*/*H* characteristic** of the material, the general form of which is shown in Fig. 3.6.

As the magnetic material is cycled around the $\boldsymbol{B}/\boldsymbol{H}$ characteristic work is done on the material and this is a source of loss in a magnetic circuit; this is the **hysteresis loss.** Referring to Fig. 3.7, for the ring of magnetic material using Ampère's law:

$$i.\mathrm{N}=\boldsymbol{H}.\boldsymbol{l} \tag{3.13}$$

where $\boldsymbol{l}$ is the path length in the material. From Faraday's law

$$v=N.\frac{\mathrm{d}\boldsymbol{\Phi}}{\mathrm{d}t}=NA.\frac{\mathrm{d}\boldsymbol{B}}{\mathrm{d}t} \tag{3.14}$$

Hence

$$v.i=A.\boldsymbol{l}.\boldsymbol{H}.\frac{\mathrm{d}\boldsymbol{B}}{\mathrm{d}t} \tag{3.15}$$

when

$$\int v.i.\mathrm{d}t=A.\boldsymbol{l}\int \boldsymbol{H}.\mathrm{d}\boldsymbol{B} \tag{3.16}$$

Hence

$$\text{Energy loss/cycle} \propto \text{Area of } \boldsymbol{B}/\boldsymbol{H} \text{ characteristic} \tag{3.17}$$

or

$$\text{Power loss/unit volume}=P_{\mathrm{h}}=\mathrm{K}_{\mathrm{h}}.f.\hat{B}^{\chi} \tag{3.18}$$

where χ is the Steinmetz coefficient and usually lies in the range 1.5 to 2; f is the frequency and K_{h} is a constant.

Comment

The Steinmetz coefficient is normally determined by experiment for a particular material.

For **electric machine** and **transformer** applications, the two most important forms of B/H characteristic are those for **low loss** or **soft** and **permanent-magnet** or **hard** nagnetic materials, respectively.

Soft magnetic materials are characterized by a narrow $\boldsymbol{B/H}$ characteristic as in Fig. .8 with low **coercivity** and are used to form the cores of electric machines and ransformers.

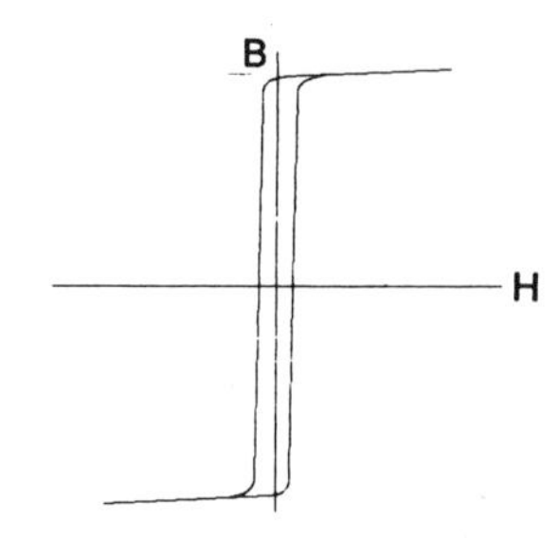

Fig. 3.8 B/H characteristic for a soft magnetic material.

Hard magnetic materials are characterized by a **high residual flux density** and a high oercivity as in Fig. 3.9. To create a permanent-magnet the material would be nagnetized by driving it into the saturation region of the $\boldsymbol{B/H}$ characteristic. The actual perating point would, however, be determined by the application of a demagnetizing orce greater than that expected in use. Operation of the permanent-magnet would then e on the recoil path shown. This will be sensibly linear and parallel to the upper or aturation region of the $\boldsymbol{B/H}$ characteristic for common permanent-magnet materials.

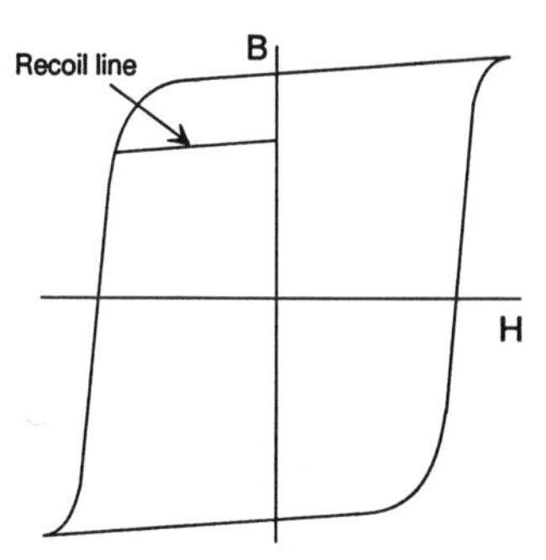

Fig. 3.9 B/H characteristic of a hard magnetic material.

Ferrite magnets find application in a range of small, low cost motors while for higher owers **alnico**, **samarium–cobalt** and, more recently, **neodymium–iron–boron** magnets re used to provide the working field of the machine.

Example 3.2 Permanent magnet

magnetic circuit consists of a permanent magnet of length 6 cm mounted between haped pole pieces of relative permeability 1120 with an effective length of 15 cm and a ross-section of 1.32 cm^2. The airgap has an effective length of 1.8 cm and a cross ection of 1.44 cm^2. Given that the appropriate portion of the hysteresis loop for the ermanent magnet material is as shown in Fig. 3.10 and that a demagnetizing field of .3 × 10^5 A T/m has been applied to the magnet, find the magnetic flux density in the irgap.

The recoil path for the magnet is parallel to the saturated region of the hysteresis oop.

For the magnetic circuit of the magnet:

$$0 = \boldsymbol{H}_{\mathrm{mag}} \cdot \boldsymbol{l}_{\mathrm{mag}} + \boldsymbol{H}_{\mathrm{mm}} \cdot \boldsymbol{l}_{\mathrm{mm}} + \boldsymbol{H}_{\mathrm{gap}} \cdot \boldsymbol{l}_{\mathrm{gap}}$$

nd

$$\boldsymbol{\Phi} = \boldsymbol{B}_{\mathrm{gap}} \cdot A_{\mathrm{gap}} = \boldsymbol{B}_{\mathrm{mm}} \cdot A_{\mathrm{mm}} = \mu_{\mathrm{o}} \cdot \mu_{\mathrm{r}} \cdot \boldsymbol{H}_{\mathrm{mm}} \cdot A_{\mathrm{mm}}$$

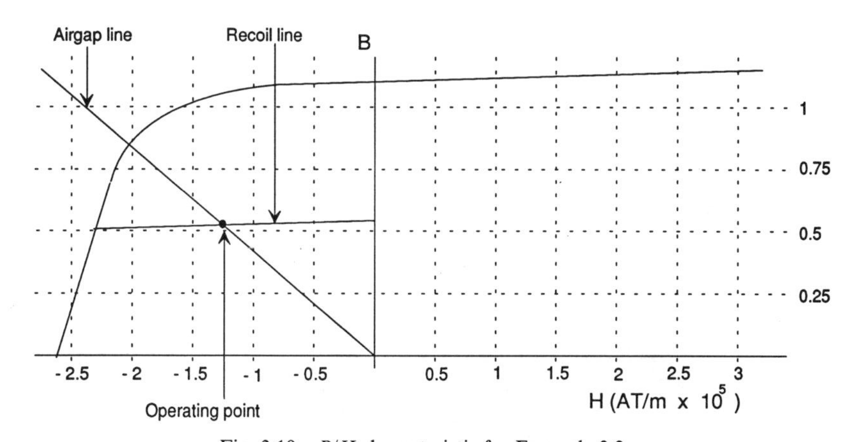

Fig. 3.10 B/H characteristic for Example 3.2.

Therefore

$$\boldsymbol{H}_{\text{mm}} = \frac{\boldsymbol{B}_{\text{gap}} \cdot A_{\text{gap}}}{\mu_{\text{o}} \cdot \mu_{\text{r}} \cdot A_{\text{mm}}}$$

when

$$\boldsymbol{H}_{\text{mag}} = -\left(\frac{\boldsymbol{B}_{\text{gap}}}{\mu_{\text{o}}}\right) \cdot \left(\frac{\dfrac{A_{\text{gap}} \cdot \boldsymbol{l}_{\text{mm}}}{\mu_{\text{r}} \cdot A_{\text{mm}}} + \boldsymbol{l}_{\text{gap}}}{\boldsymbol{l}_{\text{mag}}}\right) = -2.404 \times 10^5 \times \boldsymbol{B}_{\text{gap}}$$

The relationship $\boldsymbol{H}' = -2.404 \times 10^5 \times \boldsymbol{B}_{\text{gap}}$ is now plotted on the same axis as the hysteresis loop for the magnetic material when the intersection with the recoil path gives the operating point. From the curve the magnetic flux density in the airgap is then:

$$\boldsymbol{B}_{\text{gap}} = 0.52 \text{ T}$$

3.3.4 *Eddy current loss*

The majority of magnetic materials used in electric machines and transformers are also electrical conductors and will have a voltage and hence a current induced in them by a time varying magnetic field in line with Faraday's law.

Referring to Fig. 3.11, consider two elements of width Δx, a distance x from the centre of the material. The flux enclosed per metre of this elemental loop when subjected to a time varying magnetic field of the form $\boldsymbol{B} = \hat{B}.\sin \omega t$ is:

$$\boldsymbol{\Phi} = 2 . x . \hat{B} . \sin \ \omega t \tag{3.19}$$

The induced voltage around this loop, ignoring end effects, is then:

$$v = \frac{\mathrm{d}(2 . x . \hat{B} . \sin \omega t)}{\mathrm{d}t} = 2 . \omega . x . \hat{B} . \cos \omega t \tag{3.20}$$

As the resistance per metre of the elemental loop is $(2\rho/\Delta x)$, the induced current in the loop is

$$i_{\text{ec}} = \omega . x . \hat{B} . \cos \omega t \ \frac{\Delta x}{\rho} \tag{3.21}$$

The r.m.s. value of which is

$$I_{\text{ec}} = \omega . x . \hat{B} . \frac{\Delta x}{\sqrt{2}\, \rho} \tag{3.22}$$

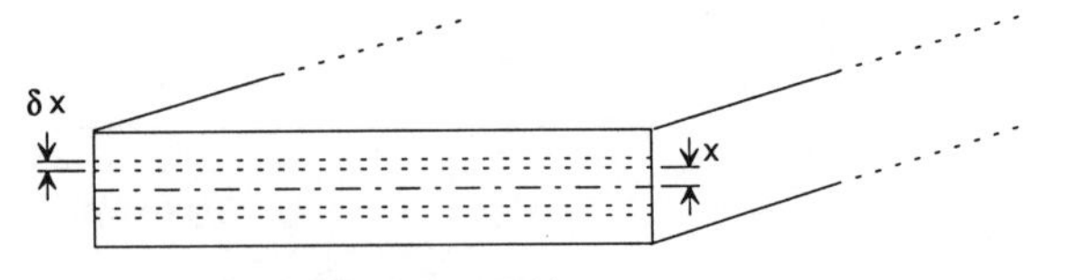

Fig. 3.11 Eddy currents.

The power loss per metre of the element is then

$$\Delta P_{ec} = I_{ec}^2 \cdot \frac{2\rho}{\Delta x} = \left(\omega \,.\, x \,.\, \hat{B} \,.\, \frac{\Delta x}{\sqrt{2}\,\rho}\right)^2 \cdot \frac{2\rho}{\Delta x} = \frac{\Delta x \,.\, (\omega \,.\, x \,.\, \hat{B})^2}{\rho} \tag{3.23}$$

The total loss per metre of the lamination is

$$P_{ec} = \int_0^{\frac{d}{2}} \frac{(\omega \,.\, x \,.\, \hat{B})^2}{\rho}\,\mathrm{d}x = \frac{\omega^2 \,.\, \hat{B}^2 \,.\, d^3}{24\rho}\ \mathrm{W/m^2} \tag{3.24}$$

where d is the thickness of the lamination, or

$$P_{ec} = \frac{\omega^2 \,.\, \hat{B}^2 \,.\, d^2}{24\rho}\ \text{watts per unit volume} \tag{3.25}$$

To reduce eddy current losses, magnetic circuits are often built up as a stack of thin laminations, each lamination being separated from its neighbour by a layer of insulation deposited on the surface of the lamination.

3.4 Self and mutual inductance

Self inductance relates the magnetic flux produced by a coil to the current in that coil according to the relationship:

$$L = \frac{N \,.\, \boldsymbol{\Phi}}{i} = \frac{\lambda}{i} = \frac{N^2}{R} \tag{3.26}$$

where $\lambda = N \,.\, \boldsymbol{\Phi}$ is the **effective flux linkages** and R is the reluctance of the magnetic ircuit.

Mutual inductance expresses the magnetic coupling between pairs of coils according o the relationship:

$$M = \frac{N_1 \,.\, \boldsymbol{\Phi}_{12}}{i_2} = \frac{N_2 \,.\, \boldsymbol{\Phi}_{21}}{i_1} \tag{3.27}$$

where

i_1 is the current in coil 1;
N_1 is the number of turns on coil 1;
$\boldsymbol{\Phi}_{12}$ is the magnetic flux linking coil 1 as a result of the current in coil 2;
i_2 is the current in coil 2;
N_2 is the number of turns on coil 2;
$\boldsymbol{\Phi}_{21}$ is the magnetic flux linking coil 2 as a result of the current in coil 1.

he direction of the mutual induced voltage in pairs of coupled coils is represented by the **dot convention** which is expressed and illustrated by Fig. 3.12.

5 Transformers

In an a.c. power system power is transmitted at high voltages – typically from 3300 volts (line) to 750 000 volts (line) – to minimize the current, and hence the I^2R loss, in

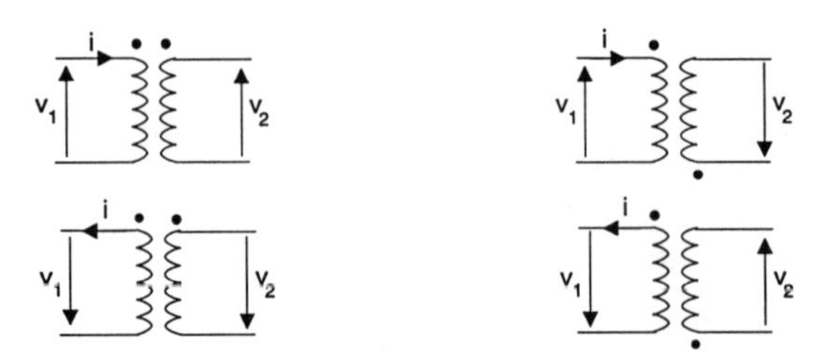

Fig. 3.12 The dot convention.

the transmission system. The majority of users, however, draw power at much lowe voltage levels of the order of 350 to 415 volts (line) (200 to 240 volts (phase)) while larg alternators generate at around 20 000 volts (line). Power transformers are used t provide the link between these various operating voltage levels.

3.5.1 *The ideal transformer*

The **ideal transformer** of Fig. 3.13 is a lossless, zero-flux device which acts as **ratio-changer** for voltage according to the relationship:

$$\frac{V_1}{V_2}=\frac{N_1}{N_2} \tag{3.28}$$

and, since the ideal transformer is a zero-flux device, ampere-turns balance exis between the two windings, then:

$$I_1 . N_1 = I_2 . N_2 \tag{3.29}$$

$$\frac{I_1}{I_2}=\frac{N_2}{N_1} \tag{3.30}$$

Also, as the ideal transformer is a lossless device:

$$\text{Power in} = \text{Power out} \tag{3.31}$$

Comment

- A **step-up transformer** is one which receives power at a low voltage and delivers it at higher voltage.
- A **step-down transformer** is one which receives power at a high voltage and delivers at a lower voltage.
- A **one-to-one transformer** receives and delivers power at the same voltage and generally used for the purpose of providing isolation.

(a) *Transfer of impedance across an ideal transformer*

Referring to Fig. 3.13, the load impedance $\boldsymbol{Z}_L$ can be expressed in terms of t secondary current and voltage as:

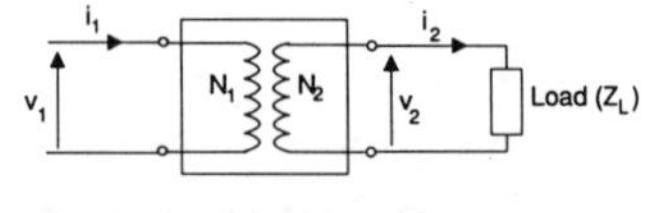

Fig. 3.13 Ideal transformer.

$$\boldsymbol{Z}_L=\frac{\boldsymbol{V}_2}{\boldsymbol{I}_2} \tag{3.32}$$

On the primary of the ideal transformer, the perceived impedance is:

$$Z'_L = \frac{V_1}{I_1} \tag{3.33}$$

Now, from equations (3.28), (3.30), (3.32) and (3.33):

$$Z_L = V_1 \cdot \frac{N_2}{N_1} \cdot \frac{N_2}{N_1} \cdot \frac{1}{I_1} = \left(\frac{V_1}{I_1}\right) \cdot \left(\frac{N_2}{N_1}\right)^2 = Z'_L \cdot \left(\frac{N_2}{N_1}\right)^2 \tag{3.34}$$

When:

$$Z'_L = Z_L \cdot \left(\frac{N_1}{N_2}\right)^2 \tag{3.35}$$

Comment

The terms primary and secondary are used to refer to the energized or input side of the transformer and the output or load side of the transformer, respectively, and make no reference to the voltages applied or delivered.

Example 3.3 Transfer of impedance
An ideal transformer is rated at 6350 volts on its primary and 240 volts on its secondary. If a load of (3+j) ohms is connected to the secondary winding, find the effective impedance referred to the primary of the transformer and use this to calculate the currents in primary and secondary windings of the transformer.

$$Z'_L = Z_L \cdot \left(\frac{N_1}{N_2}\right)^2 = (3+\mathrm{j}) \cdot \left(\frac{6350}{240}\right)^2 = (2100+700\mathrm{j}) = 2214\underline{|18.43^\circ}\ \Omega$$

Primary current

$$I_p = \frac{6350}{2214\underline{|18.43^\circ}} = 2.87\underline{|-18.43^\circ}\ \mathrm{A}$$

Secondary current is then

$$I_s = I_p \cdot \frac{6350}{240} = 75.89\underline{|-18.43^\circ}\ \mathrm{A}$$

Check using secondary values

$$I_S = \frac{240}{(3+\mathrm{j})} = 75.89\underline{|-18.43^\circ}\ \mathrm{A}$$

3.5.2 *The practical transformer*

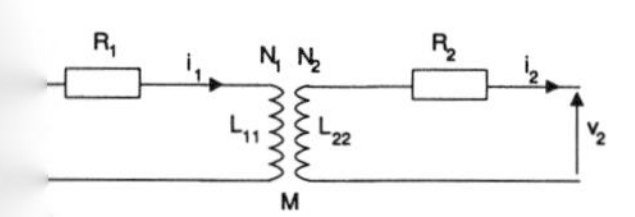

. 3.14 Single phase
nsformer circuit diagram.

Consider the circuit arrangement for single-phase transformer shown in Fig. 3.14. The equations for the primary and secondary winding circuits of this transformer are:

$$v_1 = i_1 \cdot R_1 + \frac{\mathrm{d}\lambda_1}{\mathrm{d}t} \tag{3.36}$$

and

$$0 = i_2 . R_2 + \frac{d\lambda_2}{dt} + v_2 \qquad (3.37)$$

In terms of the self and mutual inductances:

$$v_1 = i_1 . R_1 + L_{11} . \frac{di_1}{dt} - M . \frac{di_2}{dt} \qquad (3.38)$$

and

$$0 = i_2 . R_2 + L_{22} . \frac{di_2}{dt} - M . \frac{di_1}{dt} + v_2 \qquad (3.39)$$

The total flux $\boldsymbol{\Phi}_{11}$ produced by the primary winding is then related to the primary winding inductance (L_{11}) by the relationship:

$$L_{11} = \frac{N_1 . \boldsymbol{\Phi}_{11}}{i_1} \qquad (3.40)$$

However, not all the flux produced by a winding links with the opposite winding of the transformer. This non-linking flux is referred to as the leakage flux and is represented by a flux $\boldsymbol{\Phi}_1$ for the primary windings. Hence:

$$L_1 = \frac{N_1 . \boldsymbol{\Phi}_1}{i_1} \qquad (3.41$$

where L_1 is the primary leakage inductance. Similarly, for the secondary winding:

$$L_{22} = \frac{N_2 . \boldsymbol{\Phi}_{22}}{i_2} \qquad (3.42$$

and

$$L_2 = \frac{N_2 . \boldsymbol{\Phi}_2}{i_2} \qquad (3.43$$

where L_2 is the secondary leakage inductance. Now

$$M = \frac{N_1 . \boldsymbol{\Phi}_{12}}{i_2} = \frac{N_2 . \boldsymbol{\Phi}_{21}}{i_1} \qquad (3.44$$

where $\boldsymbol{\Phi}_{21}$ and $\boldsymbol{\Phi}_{12}$ are the proportions of the fluxes produced by the primary an secondary windings respectively that link with the opposite winding. Hence:

$$\boldsymbol{\Phi}_{11} = \boldsymbol{\Phi}_1 + \boldsymbol{\Phi}_{21} \qquad (3.45$$

and

$$\boldsymbol{\Phi}_{22} = \boldsymbol{\Phi}_2 + \boldsymbol{\Phi}_{12} \qquad (3.46$$

when

$$L_{11} = L_1 + M . \frac{N_1}{N_2} \qquad (3.4$$

and

$$L_{22} = L_2 + M . \frac{N_2}{N_1} \qquad (3.4$$

Substituting equations (3.47) and (3.48) in equations (3.38) and (3.39) gives:

$$v_1 = i_1 . R_1 + \left(L_1 + M . \frac{N_1}{N_2} \right) \frac{di_1}{dt} - M . \frac{di_2}{dt} \tag{3.49}$$

and

$$0 = i_2 . R_2 + \left(L_2 + M . \frac{N_2}{N_1} \right) \frac{di_2}{dt} - M . \frac{di_1}{dt} + v_2 \tag{3.50}$$

Now let

$$i_2 = i'_2 \frac{N_1}{N_2}$$

and

$$v_2 = v'_2 \frac{N_2}{N_1}$$

when

$$v_1 = i_1 . R_1 + L_1 \frac{di_1}{dt} + \frac{N_1}{N_2} . M . \left(\frac{di_1}{dt} - \frac{di'_2}{dt} \right) \tag{3.51}$$

$$0 = i'_2 \left(\frac{N_1}{N_2} \right)^2 . R_2 + L_2 \left(\frac{N_1}{N_2} \right)^2 . \frac{di'_2}{dt} + \frac{N_1}{N_2} . M . \left(\frac{di'_2}{dt} - \frac{di_1}{dt} \right) + v'_2$$

$$= i'_2 . R'_2 + L'_2 \frac{di'_2}{dt} + L_m \left(\frac{di'_2}{dt} - \frac{di_1}{dt} \right) + v'_2 \tag{3.52}$$

in which

$$R'_2 = R_2 \left(\frac{N_1}{N_2} \right)^2$$

$$L'_2 = L_2 \left(\frac{N_1}{N_2} \right)^2$$

and the magnetizing inductance

$$L_m = M \left(\frac{N_1}{N_2} \right)$$

The relationships of equations (3.51) and (3.52) are represented by the equivalent circuit of Fig. 3.15 in which the ideal transformer is included to provide the necessary ratio change.

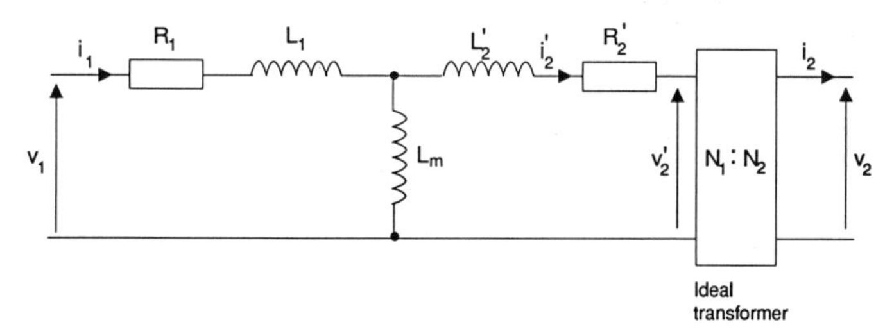

Fig. 3.15 Simple equivalent circuit of a transformer.

This equivalent circuit does not include the effect of the eddy current and hysteresis loss (**core losses**) in the magnetic material used in the transformer. These losses are a function of the transformer operating voltage and can be accommodated by the inclusion of a resistor, referred to as the **core loss resistor** (R_c), in parallel with the magnetizing inductance L_m, the combination being referred to as the **magnetizing branch** of the transformer. This gives the full equivalent circuit referred to the primary winding of Fig. 3.16.

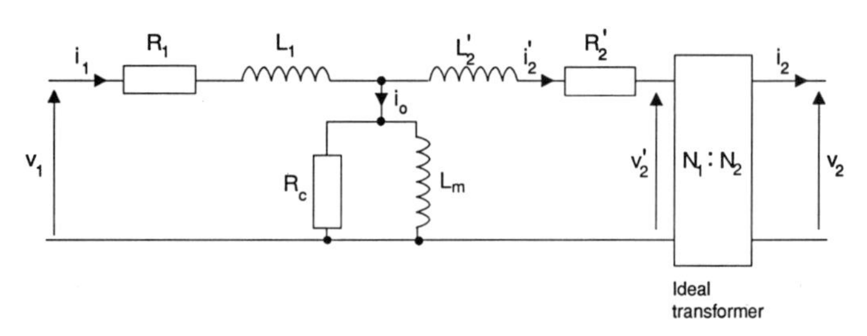

Fig. 3.16 Full equivalent circuit referred to the primary of a transformer incorporating core loss resistance.

Example 3.4 Transformer equivalent circuit (1)
A 50 Hz, 3300 volt/240 volt single phase transformer has the following parameter values:

$$R_{3300} = 18\ \Omega$$
$$X_{3300} = 60\ \Omega$$
$$R_c = 9600\ \Omega$$
$$X_m = 12\,400\ \Omega$$
$$R_{240} = 0.018\ \Omega$$
$$X_{240} = 0.098\ \Omega$$

If the transformer is supplying a voltage of 240 volts across a load of $(3+\mathrm{j})\ \Omega$, find the primary voltage that is applied to the transformer and the associated primary current.

Transfer impedances to primary

$$\boldsymbol{Z}'_L = (3+\mathrm{j}).\left(\frac{3300}{240}\right)^2 = (567.2 + 189\mathrm{j}) = 597.9\underline{|18.43^\circ}\ \Omega$$

Similarly

$$R'_{240} = 3.4\ \Omega$$

and

$$X'_{240} = 18.5\ \Omega$$

and

$$\boldsymbol{V}'_L = 3300\underline{|0^\circ}\ \mathrm{V}$$

Thus

$$I'_{\mathrm{L}} = \frac{V'_{\mathrm{L}}}{597.9\underline{|18.43^\circ}} = 5.52\underline{|-18.43^\circ} = 5.24 - 1.74\mathrm{j}\ \mathrm{A}$$

Voltage across magnetizing branch impedance is then

$$V_{\mathrm{m}} = V'_{\mathrm{L}} + I'_{\mathrm{L}}(3.4 + 18.5\mathrm{j}) = 3350.1 + 90.95\mathrm{j} = 3351.3\underline{|1.56^\circ}\ \mathrm{V}$$

Therefore

$$I_{\mathrm{o}} = \frac{V_{\mathrm{m}}}{R_{\mathrm{c}}} + \frac{V_{\mathrm{m}}}{\mathrm{j}X_{\mathrm{m}}} = 0.356 - 0.261\mathrm{j} = 0.441\underline{|-36.25^\circ}\ \mathrm{A}$$

Thus

$$I_{\mathrm{p}} = I_{\mathrm{o}} + I'_{\mathrm{L}} = 5.596 - 2\mathrm{j} = 5.94\underline{|-19.67^\circ}\ \mathrm{A}$$

Hence

$$V_{\mathrm{p}} = V_{\mathrm{m}} + I_{\mathrm{p}} . (18 + 60\mathrm{j}) = 3350.1 + 90.95\mathrm{j} + 258.2 + 350.6\mathrm{j} = 3608.3 + 441.6\mathrm{j}$$
$$= 3635.2\underline{|6.98^\circ}\ \mathrm{V}$$

(c) *Alternative forms of transformer equivalent circuit*

Figure 3.17 shows the equivalent circuit of the transformer referred to the secondary winding in which, with reference to Fig. 3.16:

$$R'_1 = R_1\left(\frac{N_2}{N_1}\right)^2$$

$$R'_c = R_c\left(\frac{N_2}{N_1}\right)^2$$

$$L'_1 = L_1\left(\frac{N_2}{N_1}\right)^2$$

nd

$$L'_{\mathrm{m}} = L_{\mathrm{m}}\left(\frac{N_2}{N_1}\right)^2$$

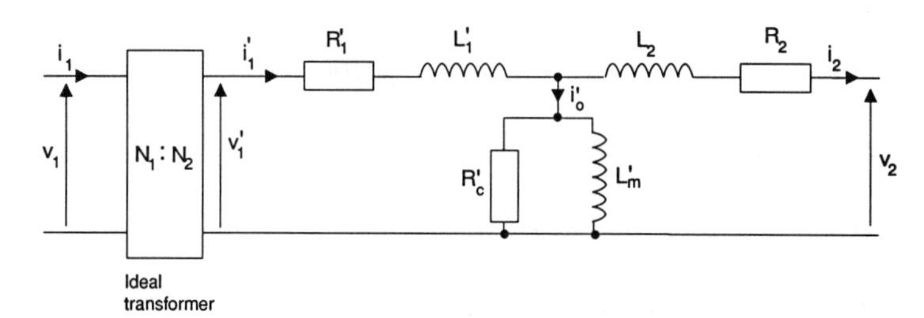

Fig. 3.17 Transformer equivalent circuit referred to secondary.

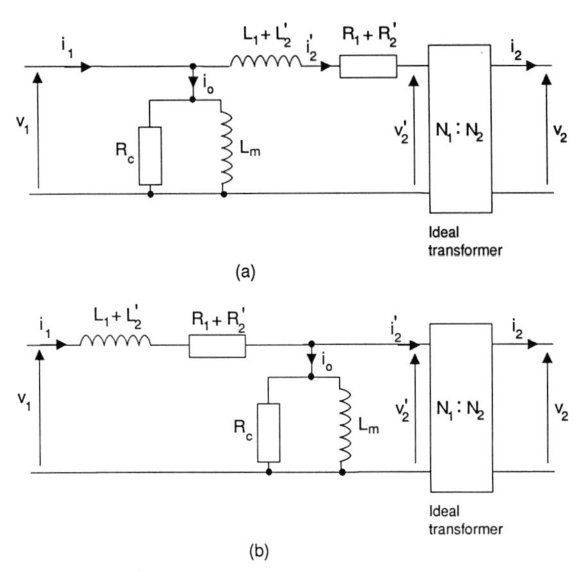

Fig. 3.18 Alternative forms of transformer equivalent circuit: (a) lumped impedance, magnetizing branch at input; and (b) lumped impedance, magnetizing branch at load.

In many instances, the analysis can be simplified by repositioning the magnetizing branch either at the input, as in Fig. 3.18(a) or, alternatively, at the load, as in Fig. 3.18(b).

Example 3.5 Transformer equivalent circuit (2)
For the transformer of Example 3.4 find the primary voltage and the current drawn from the supply assuming the approximate equivalent circuit of Fig. 3.18(a).
Load current is as before but the primary voltage is now

$$\boldsymbol{V}_p = \boldsymbol{V}'_L + \boldsymbol{I}'_L . (Z3300 + Z240') = 3300 + 5.52\angle{-18.43°} . (21.4 + 78.5j)$$
$$= 3549.1 + 373.7j = 3568.7\angle{6.01°}\ \text{V}$$

and

$$\boldsymbol{I}_p = \boldsymbol{I}'_L + \frac{\boldsymbol{V}_p}{R_c} + \frac{\boldsymbol{V}_p}{jX_m} = 5.63 - 2j = 5.97\angle{-19.56°}\ \text{A}$$

3.5.3 *Transformer testing*

The values of the equivalent circuit parameters, particularly the leakage inductances are a function of the construction of the transformer and are usually determined b mean of the **open** and **short circuit tests.**

(a) *Open circuit test*

With one winding of the transformer open circuit a voltage is applied to the othe winding such as to produce rated voltage in the open circuit winding. The applie voltage (V_1), the current (I) and the power (P_{oc}) in the energized winding are recorded as indicated by Fig. 3.19.

Referring to Fig. 3.18, as the current in the open circuit winding is zero, the current I_1 must also be zero and the current supplied may therefore be assumed to correspond to the magnetizing current I_o. Hence:

$$n = \frac{N_1}{N_2} = \frac{V_1}{V_2} \tag{3.53}$$

$$R_c = \frac{V_1^2}{P_{oc}} \tag{3.54}$$

$$I_c = \frac{V_1}{R_c} \tag{3.55}$$

$$I_m = (I^2 - I_c^2)^{\frac{1}{2}} \tag{3.56}$$

and

$$X_m = \omega L_m = \frac{V_1}{I_m} \tag{3.57}$$

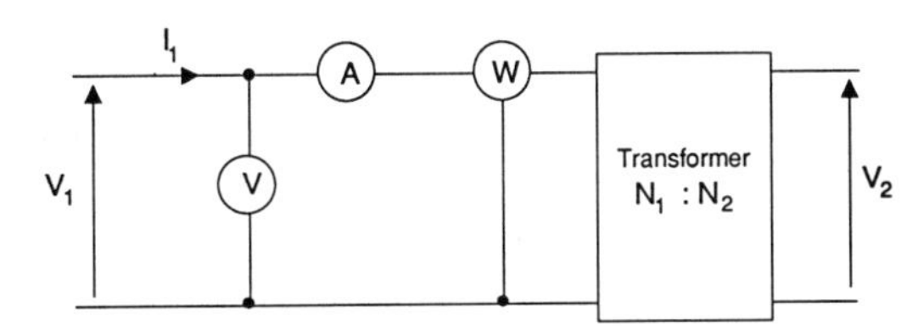

Fig. 3.19 Transformer open circuit test.

Comment

Alternatively, the voltage in the open circuit winding (V_2) can be measured in which case:

$$n = \frac{N_1}{N_2} = \frac{V_1}{V_2}$$

$$R_c = \frac{(nV_2)^2}{P_{oc}}$$

$$I_c = \frac{nV_2}{R_c}$$

$$I_m = (I^2 - I_c^2)^{\frac{1}{2}}$$

and

$$X_m = \omega L_m = \frac{nV_2}{I_m}$$

(b) *Short circuit test*

In the short circuit test, one winding of the transformer is short-circuited, and a low voltage, just sufficient to drive rated current through the short circuited winding, is applied to the other winding. This applied voltage (V_1), together with the input current (I_1) and the input power (P_{sc}) are all recorded as shown in Fig. 3.20.

As the transformer is operating at a very low voltage, typically of the order of 5% or less of the rated voltage, the effect of the magnetizing branch can be ignored. Hence:

$$Z_{\text{windings}} = \frac{V_1}{I_1} \tag{3.58}$$

$$(R_1 + R'_2) = \frac{P_{sc}}{I_1^2} \tag{3.59}$$

and

$$(X_1 + X'_2) = [Z^2_{\text{windings}} - (R_1 + R'_2)^2]^{\frac{1}{2}} \tag{3.60}$$

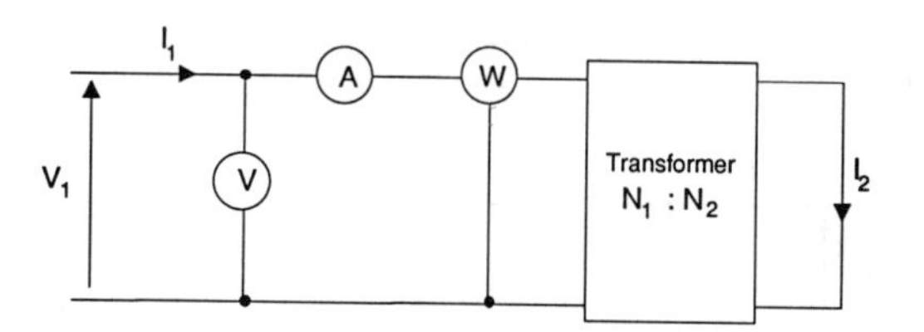

Fig. 3.20 Transformer short circuit test.

Comment

Alternatively the current in the short circuited winding (I_2) could be measured, in which case:

$$Z_{\text{windings}} = \frac{nV_1}{I_2}$$

$$(R_1 + R'_2) = P_{sc} \cdot \left(\frac{n}{I_2}\right)^2$$

and

$$(X_1 + X'_2) = [Z^2_{\text{windings}} - (R_1 + R'_2)^2]^{\frac{1}{2}}$$

Comment

The values obtained in the open and short circuit tests are each referred to the winding of the transformer that was energized when carrying out the individual tests.

Example 3.6 Open and short circuit tests

Tests on a 50 Hz, 10 000 volt/240 volt single phase transformer gave the following results:

(a) Open circuit test – low voltage winding energized

Applied voltage	240 V
Power	74 W
Current	0.98 A

(b) Short circuit test – high voltage winding energized

Applied voltage	300 V
Power	85 W
Current	0.82 A

Find the corresponding equivalent circuit values referred to the secondary winding.

From open circuit test

$$R_c = \frac{240^2}{74} = 778.4\ \Omega \qquad \text{referred to secondary}$$

Therefore

$$I_c = \frac{240}{778.4} = 0.308\ \text{A}$$

when

$$I_m = (0.98^2 - I_c^2)^{\frac{1}{2}} = 0.93\ \text{A}$$

Hence

$$X_m = \frac{240}{0.93} = 258.1\ \Omega \qquad \text{referred to secondary}$$

From short circuit test

$$(R_1 + R_2') = \frac{85}{0.82^2} = 126.4\ \Omega \qquad \text{referred to primary}$$

and

$$Z_w = \frac{300}{0.82} = 365.9\ \Omega$$

when

$$(X_1 + X_2') = (Z_w^2 - 126.4^2)^{\frac{1}{2}} = 342.4\ \Omega \qquad \text{referred to primary}$$

Referring to secondary

$$(R_1' + R_2) = 126.4\left(\frac{240}{10\,000}\right)^2 = 0.0728\ \Omega$$

and

$$(X'_1 + X_2) = 0.197\ \Omega$$

3.5.4 *Efficiency*

Efficiency is defined in terms of the power transfer as

$$\text{Efficiency} = \frac{\text{Output power}}{\text{Input power}} = \frac{\text{Output power}}{(\text{Output power} + \text{Losses})}$$

$$= \frac{\text{Output Power}}{(\text{Output power} + I^2R \text{ loss} + \text{Core loss})} \tag{3.61}$$

Of the losses, the core loss is the combination of the eddy current and hysteresis losses and is sensibly constant at constant voltage while the I^2R or **copper loss** is a function of the current in the transformer and varies with the load.

3.5.5 *Regulation*

The regulation of a transformer is expressed in terms of the volt drop on load as:

$$\text{Percentage regulation} = \frac{100\,.\,(V_{\text{no-load}} - V_{\text{load}})}{V_{\text{no-load}}} \tag{3.62}$$

3.5.6 *Three-phase transformers*

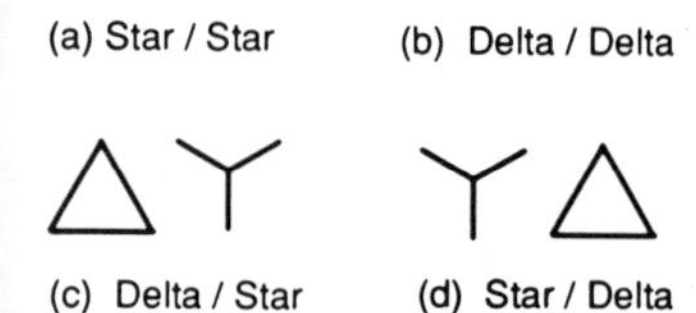

Fig. 3.21 Three-phase transformer connections: (a) star–star; (b) delta–delta; (c) delta–star; and (d) star–delta.

The possible connections for a three-phase transformer are shown in Fig. 3.21 and may be realized by the connection of three single-phase transformers as in Fig. 3.22 for the **star–delta** (Y/Δ) configuration. More usually, a three-phase transformer will be constructed in a single unit using one or other of the configurations of Fig. 3.23.

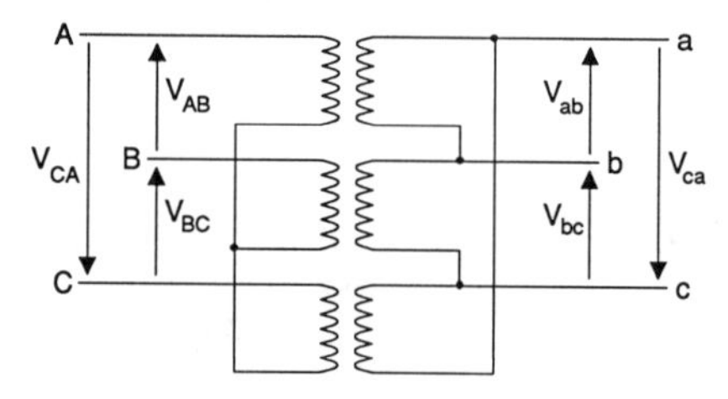

Fig. 3.22 Three-phase star–delta connected transformer.

The output voltage of both the **star–star** (Y/Y) and **delta–delta** (Δ/Δ) configurations will either be in phase with or 180° out of phase with the applied voltage. In the case of the star–delta (Y/Δ) and **delta–star** (Δ/Y) connections the output voltage will be shifted by $\pm 30°$ depending upon the relationships between connected windings.

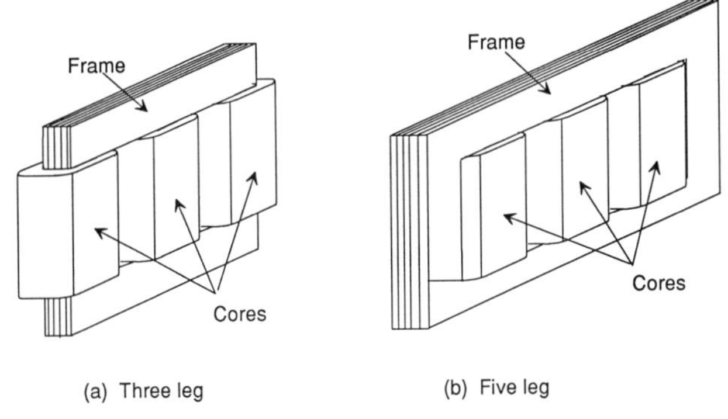

Fig. 3.23 Three-phase transformer construction: (a) three leg; and (b) five leg.

Example 3.7 Three phase transformers
A three-phase Δ/Y transformer is formed by connecting three single-phase 11 000 volt/240 volt transformers as shown in Fig. 3.24. Find the effective line-to-line turns ratio of the resulting transformer.

On the primary side of the three-phase transformer the line voltage is also the voltage across the primary of the single phase transformer.

On the secondary side of the three-phase transformer the line voltage must be $\sqrt{3}$ times the voltage across the secondary of the single-phase transformer.

The line-to-line turns ratio is then:

$$N_{3\phi} = \frac{11\,000}{240\,.\sqrt{3}} = 26.46$$

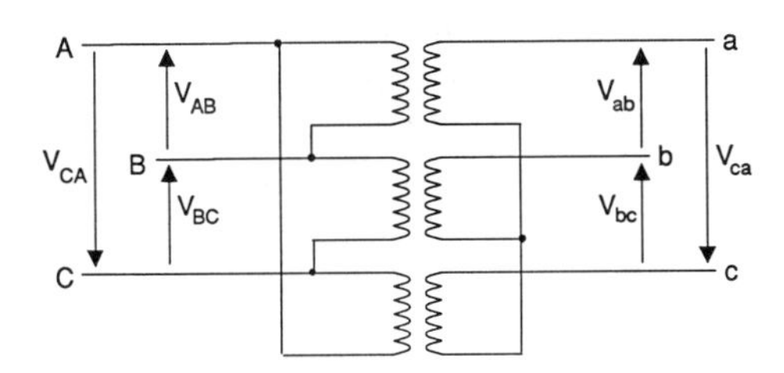

Fig. 3.24 Delta–star connected three-phase transformer.

3.5.7 *Autotransformers*

Where isolation is not a requirement an **autotransformer** can be used instead of a dual winding transformer to provide either a step-up or a step-down of voltage. The circuit diagram for an ideal autotransformer is shown in Fig. 3.25, where:

$$\frac{V_1}{V_2} = \frac{N_1}{N_2} \tag{3.63}$$

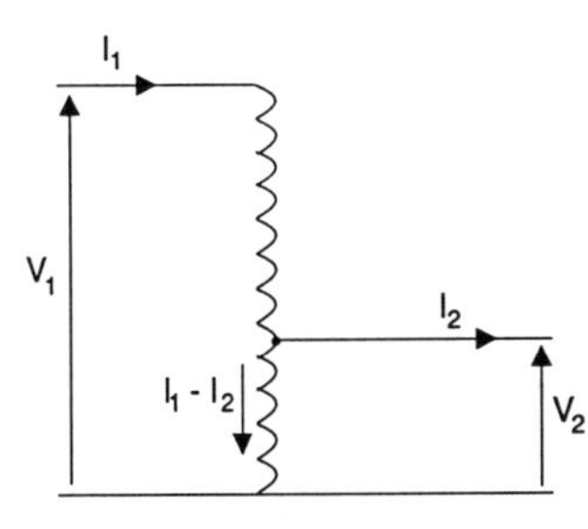

Fig. 3.25 Autotransformer.

and

$$\frac{I_1}{I_2} = \frac{N_2}{N_1} \tag{3.64}$$

The current in the lower part of the autotransformer winding is $(\boldsymbol{I}_1 - \boldsymbol{I}_2)$.

3.6 Electro-mechanical energy conversion

3.6.1 *Flux production in machines*

(a) *Direct current machines*

In a conventional d.c. machine the main field is produced at the **stator field poles** by the field windings as illustrated by Fig. 3.26 and is stationary in time and space within the machine. Torque is produced by the interaction of this field with the current carrying conductors of the roating or armature winding.

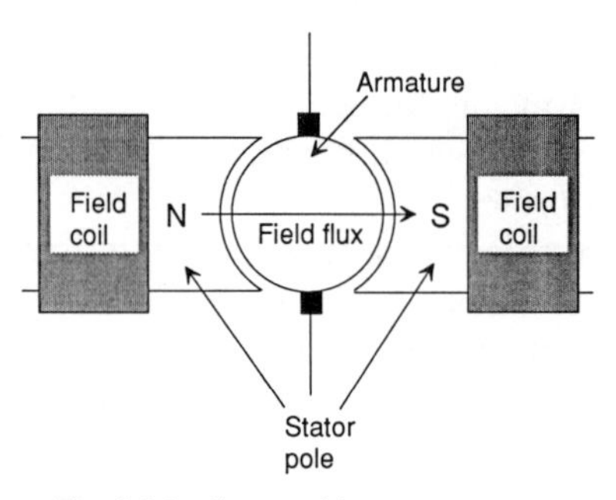

Fig. 3.26 d.c. machine construction.

(b) *Three-phase a.c. machines*

In a three-phase a.c. machine a rotating field is produced by the combination of the three-phase currents in a distributed three-phase winding.

Consider the simplified arrangement of Fig. 3.27 in which a balanced three-phase set of currents i_1, i_2 and i_3 are used to energize three coils each of N turns arranged with their axes aligned at 120° to each other. The m.m.f. vectors associated with each coil are then:

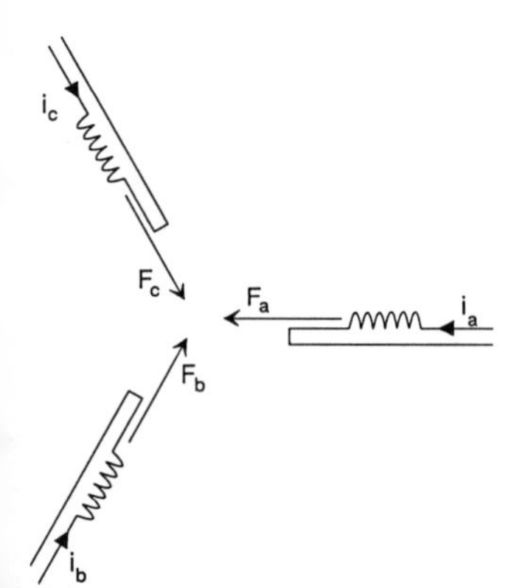

Fig. 3.27 Simplified three-phase, two-pole winding arrangement.

$$\begin{vmatrix} \boldsymbol{F}_a \\ \boldsymbol{F}_b \\ \boldsymbol{F}_c \end{vmatrix} = N.\hat{I}.\begin{vmatrix} \sin \omega t \\ \sin\left(\omega t - \frac{2\pi}{3}\right) \\ \sin\left(\omega t - \frac{4\pi}{3}\right) \end{vmatrix} \tag{3.65}$$

Taking the direction of the axis of coil A as the reference direction the reference ($\boldsymbol{F}_r$) and quadrature ($\boldsymbol{F}_q$) components of the resulting field formed by the combination of $\boldsymbol{F}_a$, $\boldsymbol{F}_b$ and $\boldsymbol{F}_c$ in space and time are:

$$\boldsymbol{F}_r = \frac{3}{2} N.\hat{I} \sin \omega t \tag{3.66}$$

and

$$F_q = \frac{3}{2} N . \hat{I} \cos \omega t \tag{3.67}$$

respectively. The resultant field has a magnitude of:

$$\hat{F}_F = (F_r^2 + F_q^2)^{\frac{1}{2}} = \frac{3}{2} N . \hat{I} \tag{3.68}$$

at an instantaneous angle θ with the axis of coil a of:

$$\theta = \tan^{-1} \frac{F_q}{F_r} = \tan^{-1}(\cot \omega t) \tag{3.69}$$

For the above analysis, corresponding to a two-pole machine, F_F is seen to be a vector of constant amplitude rotating at an angular velocity of ω. A similar result can be achieved by considering the vector summation of the instantaneous fields.

For machines with other pole numbers a three-phase winding will result in a rotating field whose angular velocity ω_F is a function of the pole number such that:

$$\omega_F = \frac{2 . \omega}{p} = \frac{\omega}{\mathrm{p}} \tag{3.70}$$

where:

ω is the angular velocity of the supply;
p is the number of poles for the machine; and
$\mathrm{p}(= p/2)$ is the number of pole pairs.

(c) *Poles and pole numbers*

An electrical machine is generally specified as having a number of poles, the presence of which influences the way in which a machine operates. In a d.c. machine, poles are generally a visible part of the structure of the machine. However, in the case of an a.c. machine, although the concept of a pole again corresponds to the distribution of the flux produced by the machine windings in the airgap of the machine, no visible physical relationship exists in the manner of a d.c. machine. Instead, the poles of an a.c. machine must be considered in terms of the distribution of the rotating magnetic field in the airgap of the machine.

For the idealized three-phase, two-pole winding considered earlier this would result in an instantaneous flux distribution in the airgap of the form shown in Fig. 3.28, the peaks of which move according to the relationships of equation (3.69). The corresponding distribution for a four-pole machine would then be as shown in Fig. 3.29.

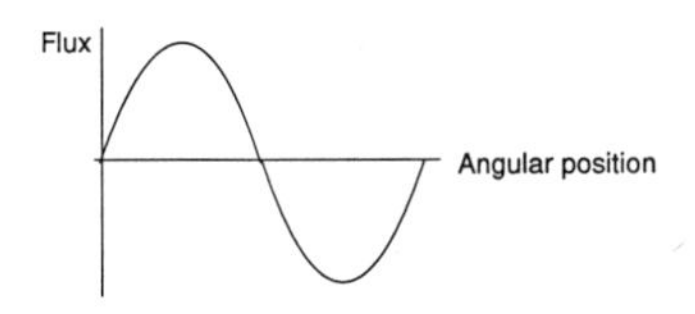

Fig. 3.28 Flux distribution in the air gap of a two-pole machine using a distributed winding of the general form of Fig. 3.27.

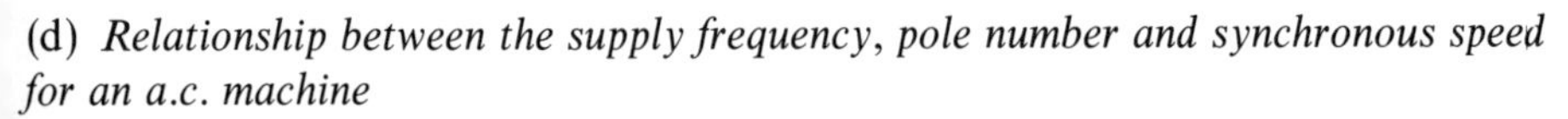

(d) *Relationship between the supply frequency, pole number and synchronous speed for an a.c. machine*

Consider now the arrangement of Fig. 3.30(a). The voltage induced in coil a would be of the form shown in Fig. 3.30(b) at a frequency determined by the speed of rotation of the rotor. If the rotor of Fig. 3.30(a) is now replaced by that of Fig. 3.31(a), then, for the same speed of rotation, the induced voltage in coil a would be at twice the frequency, as shown in Fig. 3.31(b).

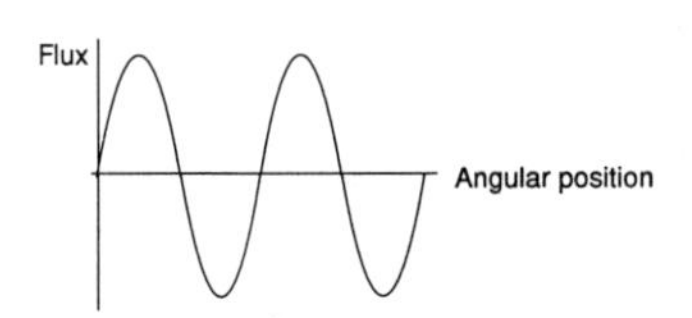

Fig. 3.29 Flux distribution in the air gap of a four-pole machine.

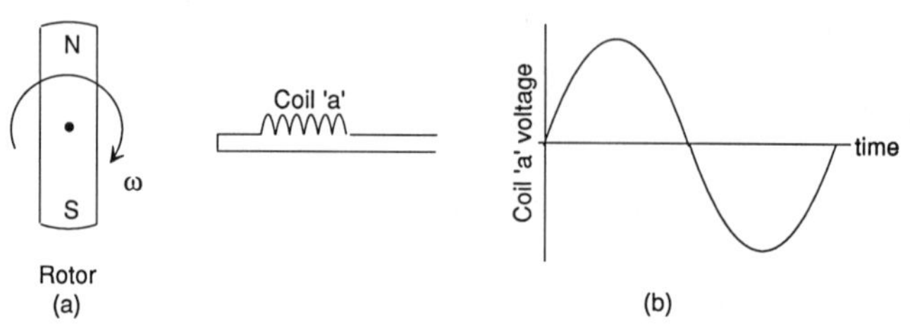

Fig. 3.30 Voltage induced in a stator coil by a two-pole rotor rotating at constant speed: (a) two-pole rotor and stator; and (b) stator coil voltage.

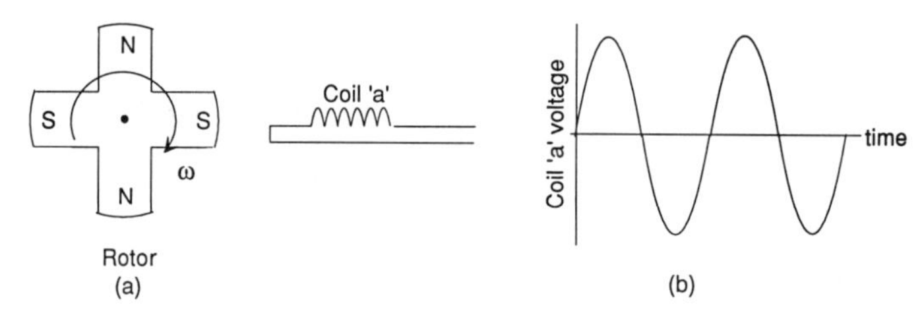

Fig. 3.31 Voltage induced in a stator coil by a four-pole rotor rotating at constant speed: (a) four-pole rotor and stator; and (b) stator coil voltage.

However, an a.c. machine would most usually be connected to a fixed frequency supply, in which case the speed of the rotor of Fig. 3.31 would have to be halved in order to obtain the same frequency at the machine terminals. Thus the synchronous speed of the a.c. machine when operated as either a motor or a generator in association with a fixed frequency supply is determined by the frequency of the supply and the number of poles on the machine such that:

$$\text{Synchronous speed} = 120\,.\frac{f}{p} \quad \text{rev/min}$$

$$= \frac{2\,.\,\omega}{p} = \frac{\omega}{\mathrm{p}} \text{ radians/second} \tag{3.71}$$

where ω, p and p are as before

(e) *Single-phase a.c. machines*

Figure 3.32 shows the field arrangement for a basic two-pole, single-phase machine. The application of a current of the form $i = \hat{I}$ sin ωt to the field coil would result in a pulsating field

$$\boldsymbol{F}_s = N\,.\,\hat{I}\,.\sin \omega t \tag{3.72}$$

along the axis of the field coil. However, as can be seen from Fig. 3.33, this pulsating field can be generated as the instantaneous vector sum of two constant amplitude fields

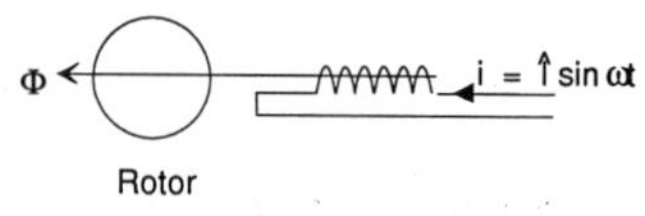

Fig. 3.32 Simplified configuration of a two-pole, single-phase machine.

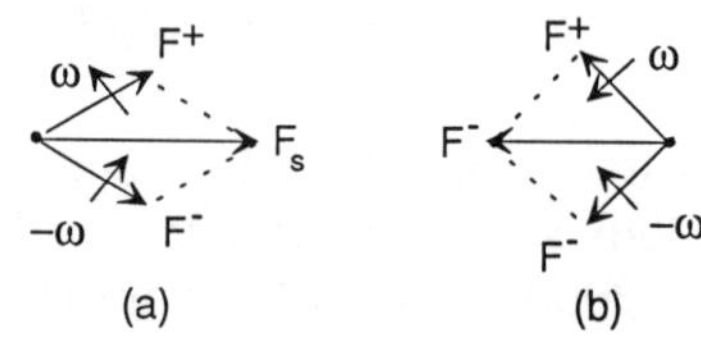

Fig. 3.33 Representation of a pulsating magnetic field by two contra-rotating fields at two arbitrary points in time: (a) t_1; (b) t_2.

$\boldsymbol{F}^+$ and $\boldsymbol{F}^-$ rotating with angular velocities $+\omega$ and $-\omega$, respectively. This production of these two, contra-rotating fields by the single field coil has implications for the operation of machines such as the single-phase induction machines described in Chapter 6.

3.6.2 *Torque production in electric machines*

As has already been stated, an electrical machine produces torque as a result of the interaction between an internal field and a set of current carrying conductors. This basic statement now needs to be expanded to a more detailed consideration of the torque production mechanism.

(a) *Singly excited systems*

Figure 3.34 shows a simple **singly excited machine** system. In operation, the torque exerted on the ferromagnetic rotor will be such as to try to bring the system to a position of minimum reluctance for the applied magnetic flux. The reluctance of the flux path is dependent upon the rotor angle and the resulting torque is referred to as **reluctance** or **saliency torque** as it results from the asymmetry or saliency of the rotor.

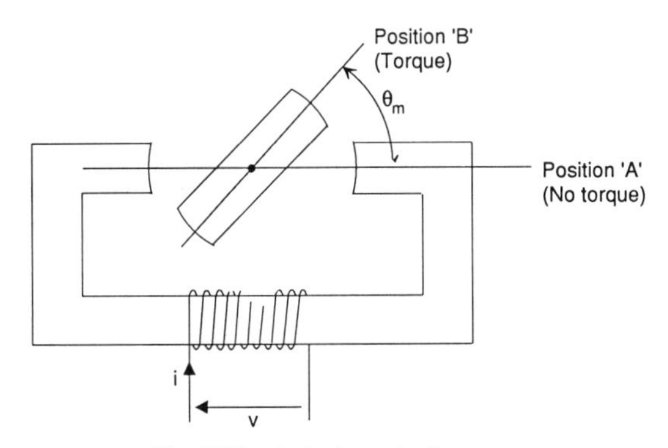

Fig. 3.34 A singly excited system.

For the system shown, ignoring mechanical losses and assuming a linear relationship between the flux linkages and current at all positions of the rotor, the voltage equation for the coil is:

$$v = i\,.\,R + L\,.\,\frac{\mathrm{d}i}{\mathrm{d}t} = i\,.\,R + N\,.\,\frac{\mathrm{d}\boldsymbol{\Phi}}{\mathrm{d}t} = i\,.\,R + \frac{\mathrm{d}\lambda}{\mathrm{d}t} \tag{3.73}$$

The instantaneous power in the coil is then:

$$v\,.\,i = i^2\,.\,R + i\,.\,\frac{\mathrm{d}\lambda}{\mathrm{d}t} \tag{3.74}$$

Integrating putting i and $\boldsymbol{\Phi}$ equal to zero at time $t=0$ gives:

$$\int_0^t v\,.\,i\,\mathrm{d}t = \int_0^t i^2\,.\,R\,.\,\mathrm{d}t + \int_0^\lambda i\,.\,\mathrm{d}\lambda \tag{3.75}$$

This corresponds to the energy balance equation with the form

Input electrical energy = Electrical losses + Effective electrical energy input (3.76)

or

$$W_e = W_{loss} + (W_{fld} + W_{em}) \tag{3.77}$$

in which W_{fld} is the energy supplied to the magnetic field and W_{em} is the electrical energy converted to mechanical energy. Hence

$$\int_0^{\lambda} i \,.\, d\lambda = W_{fld} + W_{em} \tag{3.78}$$

If the rotor is stationary, all the energy is supplied to the magnetic field when:

$$W_{fld} = \int_0^{\lambda} i \,.\, d\lambda = \int_0^{\lambda} \frac{\lambda}{L}\, d\lambda = \frac{\lambda^2}{2L} = \frac{(L \,.\, i)^2}{2L} = \frac{1}{2} L \,.\, i^2 \tag{3.79}$$

since

$$L = \frac{N \,.\, \Phi}{i} = \frac{\lambda}{i}$$

Referring to Fig. 3.35, the shaded area represents the energy stored in the field while the area under the curve represents the **co-energy** W'_{fld}. For a linear system only:

$$W_{fld} = W'_{fld} \tag{3.80}$$

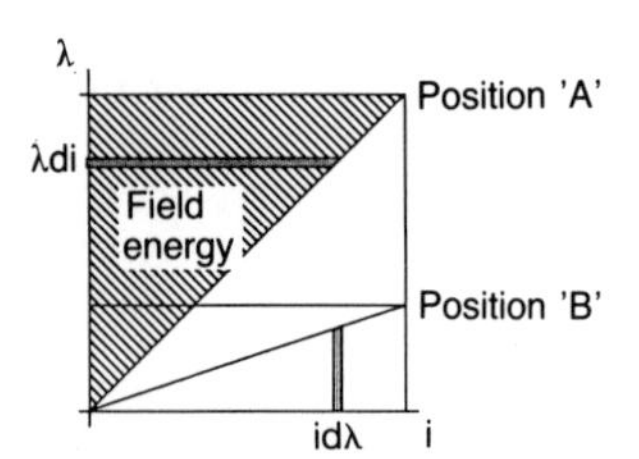

Fig. 3.35 Static flux linkage–current relationships.

Returning to Figs 3.34 and 3.35, for a small change $\Delta\theta_m$ in the rotor angle the increase in the co-energy corresponds to the mechanical work done by the system. Hence:

$$\text{Torque} = T_e = \lim_{d\theta \to 0} \left(\frac{\Delta W'_{fld}}{\Delta\theta_m}\right)_{i=\text{const}} = \left(\frac{\mathbf{d}W'_{fld}}{\mathbf{d}\theta_m}\right)_{i=\text{const}} = \left(\frac{\frac{1}{2}\,\mathbf{d}Li^2}{\mathbf{d}\theta_m}\right)_{i=\text{const}} \tag{3.81}$$

As the value of the inductance L is independent of current the expression for torque becomes that of equation (3.82).

$$T_e = \frac{i^2}{2} \cdot \frac{dL}{d\theta_m} \tag{3.82}$$

For the system of Fig. 3.34 the inductance varies with angular position according to the relationship:

$$L = L_{ave} + L_{pk} \,.\, \cos(2\theta_m) \tag{3.83}$$

where

$$L_{ave} = \frac{(L_{max} + L_{min})}{2}$$

and

$$L_{pk} = \frac{(L_{max} - L_{min})}{2}$$

Therefore

$$T_e = -i^2 \,.\, L_{pk} \,.\, \sin(2\theta_m) \tag{3.84}$$

When, for a current of the form $i = \hat{I} \sin \omega t$:

$$T_e = -\left(\frac{\hat{I}^2}{2}\right) . L_{pk} \,.\, (1 - \cos 2\omega t) \,.\, \sin(2\theta_m) \tag{3.85}$$

Assuming the rotor is rotating with an angular velocity of ω_r and that the rotor angle at time $t=0$ is $-\delta$, then:

$$\theta_m=\omega_r t-\delta \quad (3.86)$$

when

$$T_e=-\left(\frac{\hat{I}^2}{2}\right).L_{pk}.\{\sin[2(\omega_r t-\delta)]-\{\sin[2(\omega_r+\omega)t-2\delta]\}+\sin[2(\omega_r-\omega)t-2\delta]\} \quad (3.87)$$

Equation (3.87) has an average value of zero unless $\omega_r=\omega$, when:

$$T_{e,ave}=-\left(\frac{\hat{I}^2}{2}\right).L_{pk}.\sin(-2\delta)=K_e.\sin(2\delta) \quad (3.88)$$

This is the reluctance or saliency torque and arises as a result of the magnetic asymmetry of the machine.

(b) *Doubly excited systems*

Figure 3.36 shows a simple doubly excited system incorporating saliency on both rotor and stator. The flux linkage for the respective windings are then:

$$\lambda_1=L_1.i_1+M.i_2 \quad (3.89)$$

and

$$\lambda_2=L_2.i_2+M.i_1 \quad (3.90)$$

where M is the mutual inductance between the two coils.

The coil voltages are then:

$$v_1=i_1.R_1+\frac{d\lambda_1}{dt} \quad (3.91)$$

and

$$v_2=i_2.R_2+\frac{d\lambda_2}{dt} \quad (3.92)$$

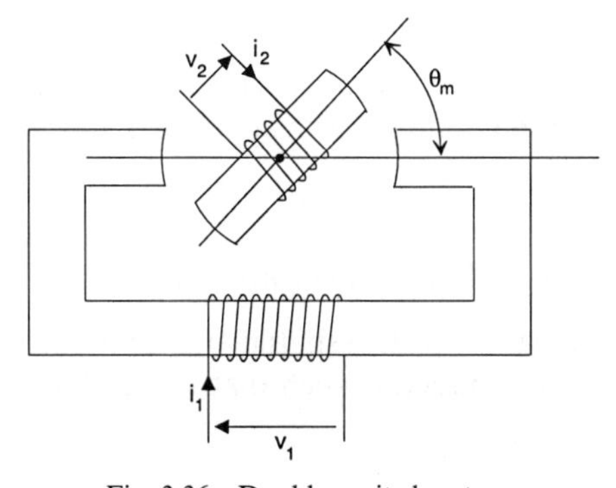

Fig. 3.36 Doubly excited system.

Now, since the self and mutual inductances are a function of rotor position and assuming currents to be functions of time and not of inductance, the coil powers are:

$$v_1 . i_1 = R_1 . i_1^2 + i_1 . \frac{d\lambda_1}{dt} = R_1 . i_1^2 + i_1 . L_1 . \frac{di_1}{dt} + i_1^2 . \frac{dL_1}{dt} + M . i_1 . \frac{di_2}{dt} + i_1 . i_2 . \frac{dM}{dt} \tag{3.93}$$

and

$$v_2 . i_2 = R_2 . i_2^2 + i_2 . \frac{d\lambda_2}{dt} = R_2 . i_2^2 + i_2 . L_2 . \frac{di_2}{dt} + i_2^2 . \frac{dL_2}{dt} + M . i_2 . \frac{di_2}{dt} + i_1 . i_2 . \frac{dM}{dt} \tag{3.94}$$

Combining equations (3.93) and (3.94) and integrating as before with i and $\boldsymbol{\Phi}$ equal to zero at time $t=0$ give the energy balance equation:

$$W_e = W_{loss} + W_{fld} + W_{em} \tag{3.95}$$

in which:

$$\text{Electrical energy input} = W_e = \int_0^t (v_1 . i_1 + v_2 . i_2) . dt$$

$$\text{Electrical losses} = W_{loss} = \int_0^t (R_1 . i_1^2 + R_2 . i_2^2) . dt$$

Energy to field + Electrical energy converted to mechanical energy

$$= W_{fld} + W_{em} = \int_0^{\lambda_1} i_1 . d\lambda_1 + \int_0^{\lambda_2} i_2 . d\lambda_2$$

Assuming a stationary rotor the energy stored in the magnetic field is:

$$W_{fld} = \frac{1}{2} L_1 . i_1^2 + \frac{1}{2} L_2 . i_2^2 + M . i_1 . i_2 \tag{3.96}$$

The change in the stored energy in the field with time as the rotor rotates can now be found when it can be shown that:

$$W_{em} = \int \left(\frac{1}{2} i_1^2 . dL_1 + \frac{1}{2} i_2^2 . dL_2 + i_1 . i_2 . dM \right) \tag{3.97}$$

Differentiating with respect to θ_m enables the expression for torque to be found when

$$T_e = \frac{1}{2} i_1^2 . \frac{dL_1}{d\theta_m} + \frac{1}{2} i_2^2 . \frac{dL_2}{d\theta_m} + i_1 . i_2 . \frac{dM}{d\theta_m} \tag{3.98}$$

In equation (3.95) the first two terms are saliency or reluctance torques while the third term is the **co-alignment torque** acting to bring the stator and rotor fields into alignment along a common axis.

(c) *Torque as the interaction of stator and rotor fields in a doubly excited machine*

For the basic machine of Fig. 3.36 with a spatial distribution of m.m.f. in the airgap due to the current in the stator conductors such that the field at some angle θ due to the stator is $\boldsymbol{F}_s(\theta)$, where:

$$\boldsymbol{F}_s(\theta) = \hat{\boldsymbol{F}}_s . \sin\theta \tag{3.99}$$

Comment

This assumes a spatial current distribution in the stator winding of the form $I_s(\theta)=\hat{I}_s \cos\theta$ since

$$F_s(\theta)=\int_0^\theta I_s(\theta)\,.\,d\theta$$

Similarly the spatial m.m.f. due to the rotor currents is:

$$F_r(\theta)=\hat{F}_r\,.\,\sin(\theta+\delta) \tag{3.100}$$

Comment

This assumes a spatial current distribution in the rotor winding of the form:

$$I_r(\theta)=\hat{I}_r\,.\,\cos(\theta+\delta).$$

The resultant spatial distribution of m.m.f. in the airgap is then:

$$F_g(\theta)=F_s(\theta)+F_r(\theta)=\hat{F}_g\,.\,\sin(\theta+\delta-\gamma) \tag{3.101}$$

in which the relationships between F_s, F_r and F_g and between γ and δ are as shown by the vector diagram of Fig. 3.37.

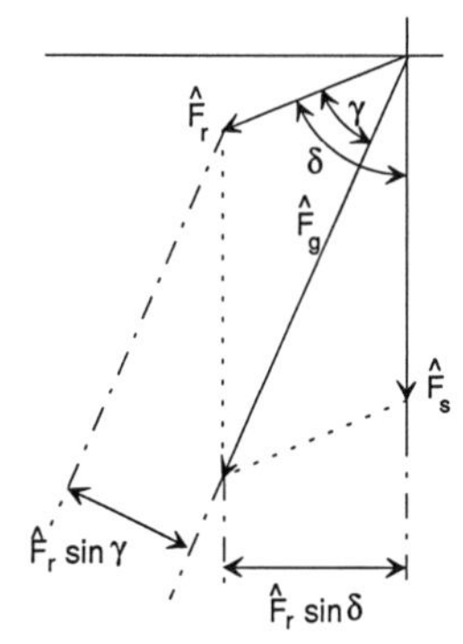

Fig. 3.37 Field vectors for a doubly excited system with both stator and rotor energized.

Assuming a zero reluctance path in the magnetic material the radial flux density in the airgap is then:

$$B_g(\theta)=\mu_o\,.\,H_g(\theta)=\mu_o\,.\,\frac{F_g(\theta)}{l_g}=\mu_o\,.\,\frac{\hat{F}_g}{l_g}\,.\,\sin(\theta+\delta-\gamma)=\hat{B}_g\,.\,\sin(\theta+\delta-\gamma) \tag{3.102}$$

where l_g is the length of the airgap.

Assuming the spatial distribution of current in the rotor referred to above the force (F) on the rotor is then, from the vector product $F=I\times B$:

$$F=l_r\,.\int_0^{2\pi}[\hat{B}_g\,.\,\sin(\theta+\delta-\gamma)\,.\,I_r\,.\,\cos(\theta+\delta-\gamma)]\,.\,d\theta=\pi\,.\,\hat{B}_g\,.\,I_r\,.\,l_r\,.\,\sin\gamma \tag{3.103}$$

where l_r is the length of the rotor of the machine.

For a rotor of radius r the torque is then:

$$T_e=\pi\,.\,r\,.\,\hat{B}_g\,.\,I_r\,.\,l_r\,.\,\sin\gamma \tag{3.104}$$

or

$$T_e \alpha F_g\,.\,F_r\,.\,\sin\gamma \tag{3.105}$$

Referring to Fig. 3.37

$$\hat{F}_r\,.\,\sin\gamma=\hat{F}_s\,.\,\sin(\delta-\gamma) \tag{3.106}$$

and

$$\hat{F}_r\,.\,\sin\delta=\hat{F}_g\,.\,\sin(\delta-\gamma) \tag{3.107}$$

Comment

Equation (3.103) is the magnitude of the vector product $\boldsymbol{I} \times \boldsymbol{B}$.

Therefore

$$T_e \alpha \boldsymbol{F}_s \,.\, \boldsymbol{F}_r \,.\, \sin \delta \tag{3.108}$$

and

$$T_e \alpha \boldsymbol{F}_g \,.\, \boldsymbol{F}_s \,.\, \sin(\delta - \gamma) \tag{3.109}$$

The torque produced by a double excited machine can therefore also be considered in terms of the individual airgap fields and their angular relationships.

Comment

The result of equation (3.108) is the magnitude of the vector product of $\boldsymbol{F}_s$ and $\boldsymbol{F}_r$.

Exercises

1 A toroid with a mean circumference of 20 cm, a cross-sectional area of 2 cm^2 and made of a magnetic material with a μ_r of 2000 is cut so as to make two identical halves. A 1000 turn coils is wound on to one of the halves. If the two halves of the toroid are now separated by two pieces of material of length 2 cm, cross-sectional area 1 cm^2 and μ_r of 1000, what will be the flux in the circuit if the coil is supplied with a d.c. current of 20 mA?

2 If the two halves are now separated by two 0.5 cm long airgaps find value of d.c. coil current required in order to produce a flux density of 0.2 T in the airgap. Assume that fringing effects increase the effective area of the airgap by 15%.

3 Assuming that the two halves mate perfectly, calculate the inductance of the coil and the effective loss resistance if an a.c. current of 15 mA flows through the coil when it is connected to a 5 volt, 50 Hz supply.

4 A magnetic circuit has the specification given in Table 3.1.

Table 3.1 Specification of a magnetic circuit

Section	Effective area (cm^2)	Effective length (cm)
Iron section (a)	715	3.7
Iron section (b)	667	17.0
Iron section (c)	800	37.0
Airgap	1790	1.0

The $\boldsymbol{B}/\boldsymbol{H}$ for all of the iron sections is given by:

$\boldsymbol{B}$ (Tesla)	1.0	1.1	1.2	1.3	1.4	1.5
$\boldsymbol{H}$ (A T/m)	190	235	295	395	575	1050

Estimate the ampere-turns required to produce a flux of 0.1 Wb in the airgap.

5 It is required to increase the airgap flux density for the circuit of exercise 3.4 to 0.115 Wb without changing the effective length of any of the sections. Redesign the magnetic circuit in order to achieve the new value of airgap flux and calculate the new value of ampere-turns required.

6 A horseshoe magnet has a cross-sectional area of 10 cm^2 at each pole face. If it is designed to lift an iron plate of mass 100 kg, what will be the minimum flux density required at the pole faces?

7 Open and short circuit test on a single phase 4 kV A, 200/400 volt, 50 Hz transformer gave the following results:

Open circuit test, low voltage side energized: $V=200$ V, $I=0.7$ A and power $=70$ W
Short circuit test, high voltage side energized: $V=15$ V; $I=10$ A and power $=80$ W

Determine the equivalent circuit for the transformer and hence find the input voltage, input current and efficiency of the transformer when it is supplying rated volt-amperes at 400 volt and unity power factor.

8 A transformer equivalent circuit the form of equivalent circuit of Fig. 3.18(a) has the following parameter values:

$$\omega L_m = 231\text{j}\ \Omega$$
$$R_c = 400\ \Omega$$
$$R_1 + R'_2 = 0.16\ \Omega$$
$$\omega(L_1 + L'_2) = 0.7\text{j}\ \Omega$$
$$N_1 : N_2 = 1:10$$

For an applied primary voltage of 200 volts and a load impedance of (596 + 440j) W connected to the secondary find (a) the load voltage (b) the input current (c) the power in and (d) the power out.

9 A three-phase Δ/Y transformer has 2200 turns on each of its primary windings and 48 turns on each secondary winding. If 11 000 volts (line) is applied to the primary of the transformer, what will be the secondary line voltage?

4 The d.c. machines

In a conventional d.c. machine a stationary main field produced by the field windings on the stator of the machine interacts with a current flowing in the **rotor** or **armature** conductors to develop a torque. Consider the situation shown in Fig. 4.1(a). For the directions of stator field and armature current shown, a torque will be produced which will cause the armature to rotate in a clockwise direction. When the armature reaches the position of Fig. 4.1(b), the conductors making up coil 1 have moved out of the field to a position on the **neutral axis** and no torque is developed.

The position of Fig. 4.1(b) is in fact a stable condition as the stator field and the armature field – that produced by the armature current – are now aligned and any further clockwise movement of the armature would result in an anti-clockwise torque acting to restore the stable state.

In order to maintain the rotation, the current in coil 1 must now be turned off and a similar current introduced into coil 2, producing a further rotation of 90° after which the current in coil 2 must be turned off and that in coil 1 restored, but in the reverse direction. After a further 90° rotation the current in coil 2 is reversed after which the armature has completed a single revolution. An arrangement with only two armature coils is obviously inefficient and a practical machine has a large number of armature coils and the switching of armature coil current is arranged to ensure that all the conductors under a field pole carry current in the same direction, as in Fig. 4.1(c).

In a conventional d.c. machine the switching of the current in the armature coils is achieved mechanically by means of the **commutator**. As indicated by Fig. 4.1(d), this consists of a ring of insulated copper segments to which the ends of the armature coils are connected. Power is transferred to the armature coils by means of **brushes** which run on the surface of the commutator. The resulting field produced by the armature is similar to that which would be produced by a stationary coil of the appropriate number of turns connected between the brushes as in Fig. 4.1(e).

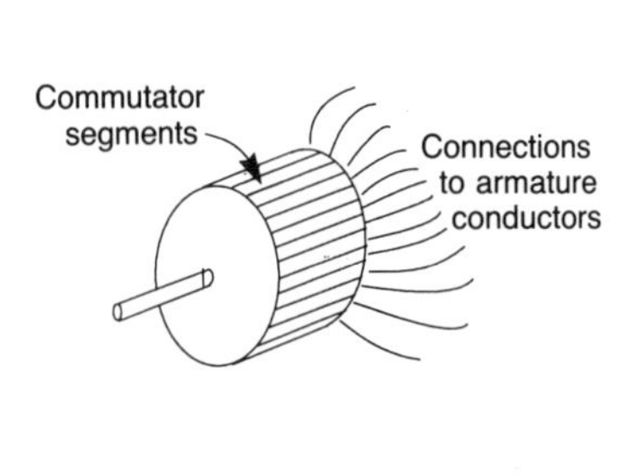

Figure 4.1 (d)

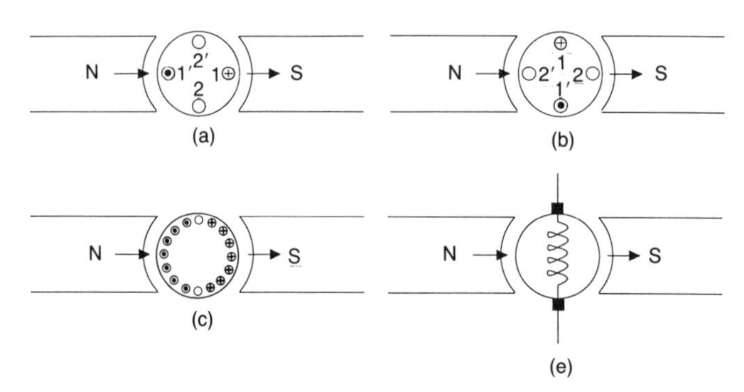

Fig. 4.1 Torque production in a d.c. machine: (a) initial position with coil 1 in field; (b) position after 90° rotation with coil 1 on neutral axis; (c) armature with multiple conductors; (d) commutator; and (e) replacement of armature conductors with a pseudo-stationary coil.

Comment

The torque produced by the d.c. machine is therefore a co-alignment torque (Chapter 3).

As was shown in Chapter 3, the movement of a conductor in a magnetic field results in the production of an e.m.f. in that conductor. In the case of the d.c. machine, although the pattern of current in the armature coils is stationary with respect to the stator, the armature conductors are moving through the stator field. This results in the production of a voltage in the armature coils referred to as the armature voltage or back e.m.f. In the case of a d.c. motor this voltage opposes the applied voltage and is the source voltage in the case of a d.c. generator.

The direction of rotation of a d.c. motor is determined by the direction of the field under the individual poles and the direction of the associated currents in the armature. If the direction of either the field or the armature current is reversed then the direction of rotation of the motor will be reversed. However, should both the field and the armature current be reversed together then the direction of rotation of the armature will remain the same.

4.1 Steady state e.m.f. and torque equations

Let the airgap flux per pole of a p pole machine be $\boldsymbol{\Phi}$. The flux density ($\boldsymbol{B}$) under a pole is then, for a rotor of length l and radius r:

$$\boldsymbol{B}=\frac{\boldsymbol{\Phi}}{A} \tag{4.1}$$

where

$$A=\frac{2\pi rl}{p}$$

is the area under a single pole. If the rotor is rotating at n rev/min, the angular velocity of a rotor conductor is then:

$$\omega_{\mathrm{rm}}=\frac{2\pi n}{60} \tag{4.2}$$

The voltage induced in a single rotor conductor is then, using:

$$e=\boldsymbol{B}\,.\,l\,.\,r\,.\,\omega_{\mathrm{rm}} \tag{4.3}$$

Combining equations (4.1) and (4.3):

$$e=\frac{p\boldsymbol{\Phi}\omega}{2\pi} \quad \text{volts/conductor} \tag{4.4}$$

or

$$e=\frac{p\boldsymbol{\Phi}n}{60} \quad \text{volts/conductor} \tag{4.5}$$

For a total of Z_α armature conductors connected in α parallel paths the total induced voltage in the armature (armature voltage or back e.m.f. E_a)is:

$$E_a = \frac{p \, . \, \boldsymbol{\Phi} \, . \, \omega_{rm} \, . \, Z_\alpha}{2\pi\alpha} \quad \text{volts} \tag{4.6}$$

or

$$E_a = \frac{p \, . \, \boldsymbol{\Phi} \, . \, n \, . \, Z_\alpha}{60\alpha} \quad \text{volts} \tag{4.7}$$

where n is the speed in rev/min.

The number of parallel paths (α) in the winding is a function of the winding configuration. The two most common windings for the armature of a d.c. machine are the **wave winding** which provides two parallel paths irrespective of the number of poles ($\alpha = 2$) and the **lap winding** in which the number of parallel paths is the same as the number of poles ($\alpha = p$).

The equations for E_a are more usually written in terms of the flux per pole and rotor speed as:

$$E_a = \mathrm{K} \, . \, \boldsymbol{\Phi} \, . \, \omega_{rm} = \mathrm{K}' \, . \, \boldsymbol{\Phi} \, . \, n \tag{4.8}$$

Under steady state conditions with the d.c. machine operating as a motor the voltage (V) applied to the machine armature is equal to the sum of the back e.m.f. (E_a) and the voltage drop across the armature resistance ($I_a R_a$). Thus:

$$V = E_a + I_a R_a \tag{4.9}$$

where I_a is the armature current.

The power supplied to the armature of the machine is also made up of two components, the copper loss in the armature conductors ($I_a^2 R_a$) and the mechanical power developed at the shaft of the machine. Referring to equation (4.9) the power supplied to the rotor is:

$$V \, . \, I_a = E_a \, . \, I_a + I_a^2 \, . \, R_a \tag{4.10}$$

The mechanical power developed at the shaft of the machine (P_{mech}) is then:

$$P_{mech} = E_a \, . \, I_a \tag{4.11}$$

or, in terms of the shaft torque (T) and the shaft speed (ω_{rm}):

$$P_{mech} = T \, . \, \omega = E_a \, . \, I_a \tag{4.12}$$

Combining equations (4.8) and (4.12) gives the expression for the shaft torque in terms of the flux per pole and the armature current.

$$T = \mathrm{K} \, . \, \boldsymbol{\Phi} \, . \, I_a \tag{4.13}$$

It is important at this point to note that the mechanical power and torque referred to in equations (4.11), (4.12) and (4.13) are the mechanical power and torque developed by the machine and that these differ from the useful mechanical power and torque at the output shaft by the mechanical losses of the machine. These mechanical losses are du to factors such as the friction in the shaft bearings and the air resistance or **windage** c the rotating armature.

The useful mechanical power output of the machine can therefore be expressed as

$$\text{Useful mechanical power} = \text{Mechanical power developed} - \text{Rotational losses} \tag{4.14}$$

4.2 Armature circuit

Fig. 4.2 d.c. machine armature current.

The armature circuit of a d.c. machine is represented by Fig. 4.2 and yields equations (4.15) and (4.16) for a d.c. machine operating at steady state in motoring and generating modes respectively.

Motoring $V=E_a+I_aR_a$ (4.15)

Generating $E_a=V+I_aR_a$ (4.16)

When motoring, the armature current flows in the direction shown in Fig. 4.2 and energy is transferred from the electrical system to the mechanical system. When generating the direction of current is reversed and energy is transferred from the mechanical system to the electrical system.

In practice there will be a small volt drop, typically of the order of 1 volt/brush, at the brushes of the d.c. machine. This brush volt drop (V_b) can be incorporated into equations (4.15) and (4.16) as follows:

Motoring $V=E_a+I_aR_a+V_b$ (4.17)

Generating $E_a=V+I_aR_a+V_b$ (4.18)

Comment

The brush volt drop (V_b) can be subsumed into the voltage at the armature terminals using the relationship:

$$V\mp V_b=V'=E_a\pm I_a\cdot R_a$$

and can therefore be considered as a part of the effective armature terminal voltage.

In many instances, the brush volt drop is small in comparison with the applied voltage and can be ignored.

.3 Construction

.3.1 *Radial flux machines*

he majority of d.c. machines in use are of the **radial flux**, **heteropolar** type in which the irection of the flux relative to the rotor conductors reverses between successive poles s in Fig. 4.1. The flux is produced by field windings carried on the stator field poles as hown in Fig. 4.3 for a four-pole machine. This configuration is referred to as **salient ole** construction. The magnitude of the field flux is a function of the field current, the umber of turns on the field winding and the reluctance of the flux path.

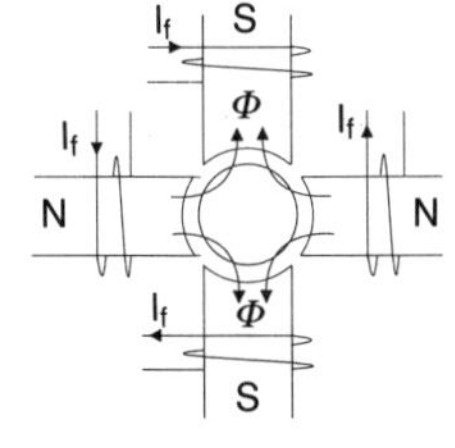

Fig. 4.3 Four-pole salient pole d.c. machine.

The armature windings are carried on the armature or rotor of the machine with the ondutors buried in a series of slots in the surface of the armature as in Fig. 4.4. The nds of the armature conductors are brought out to the commutator segments.

---

omment

he slots of Fig. 4.4 are evenly distributed around the perimeter of the armature and

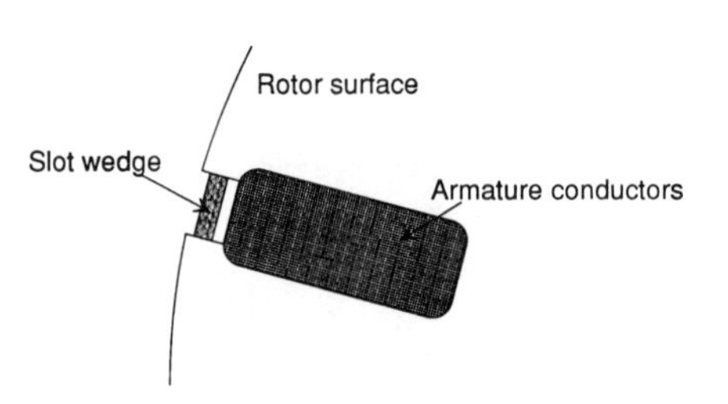

Fig. 4.4 Armature conductors in rotor slot.

the armature winding may therefore be referred to as a *distributed winding*. This is in contrast with the salient pole configuration associated with the field winding.

An alternative form of radial flux machine is the **homopolar machine** of Fig. 4.5 in which the direction of the field flux relative to the armature conductors remains constant. The main disadvantage of the homopolar machine is the relatively short active length of conductor in the field resulting in a higher armature current per unit of torque than a heteropolar machine.

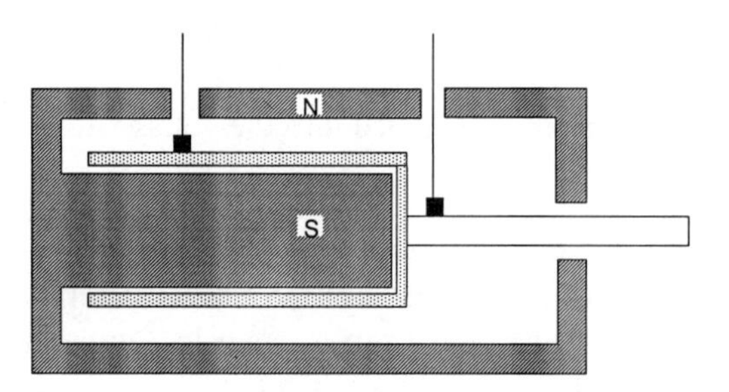

Fig. 4.5 Radial flux homopolar d.c. machine.

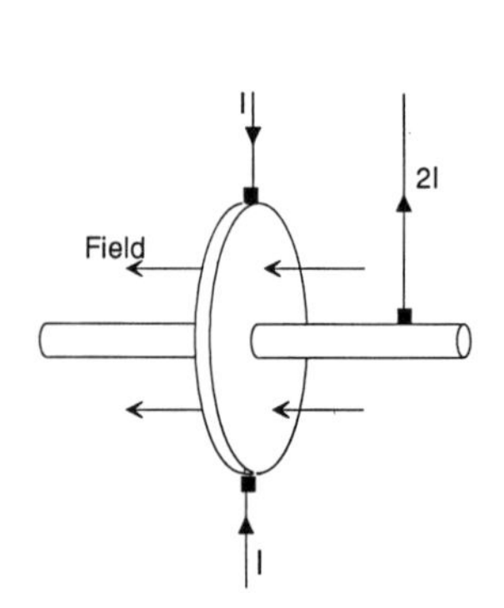

Fig. 4.6 Axial flux d.c. machine.

4.3.2 *Axial flux machines*

Both the heteropolar and homopolar machines can be constructed with an axial flu path as in Fig. 4.6. This type of construction has the advantage of providing a ver compact structure which has benefits in certain areas of application such as whe mounted motors for electric vehicles and was also used for some of the earl superconducting motors.

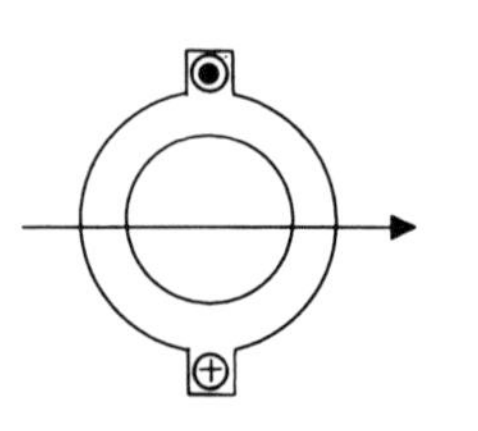

Fig. 4.7 Flux path for single stator coil.

4.4 Armature reaction

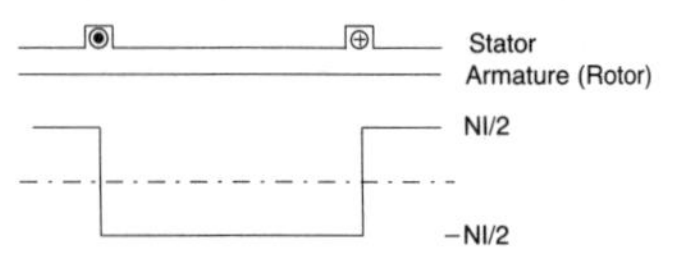

Fig. 4.8 m.m.f. diagram for a single coil.

Consider the single coil of N turns in Fig. 4.7. Referring to the work on magnet circuits in Chapter 3, as the core of the machine is constructed from a high relativ permeability magnetic material, all the reluctance of the magnetic circuit can b assumed to be concentrated in the airgap of the machine. Then if the coil is carrying current I, the m.m.f. along the path shown becomes NI with $NI/2$ appearing across eac airgap. Developing the diagram of Fig. 4.7 then gives the m.m.f. diagram of Fig. 4.

This may then be expanded to provide the m.m.f. diagram for a uniformly distributed winding such as that referred to above, which is shown in Fig. 4.9, which with a large number of coils approximates to the form of Fig. 4.10 which may be taken as representative of many armature windings. The flux distribution associated with the armature m.m.f. is shown in Fig. 4.11.

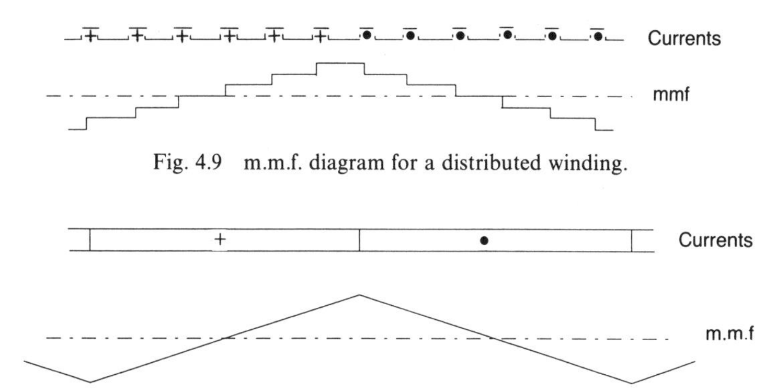

Fig. 4.9 m.m.f. diagram for a distributed winding.

Fig. 4.10 m.m.f. diagram for a distributed winding with a large number of coils.

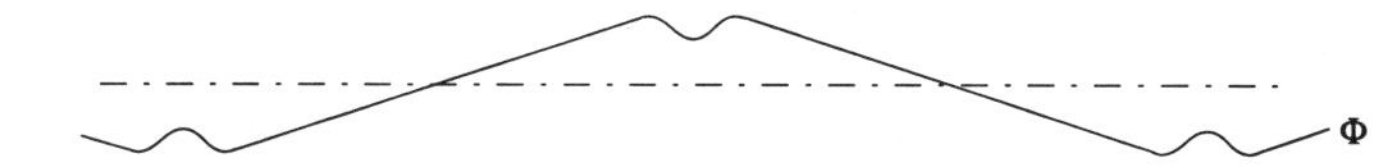

Fig. 4.11 Flux distribution produced in the air gap by the m.m.f. distribution of Fig. 4.10.

Comment

The dip in the peak flux in Fig. 4.11 results from the increase in reluctance of the flux path in the region between the field poles.

Consider now a salient pole field winding. The m.m.f. diagram and flux distribution for a winding of this type are then as shown in Figs 4.12 and 4.13.

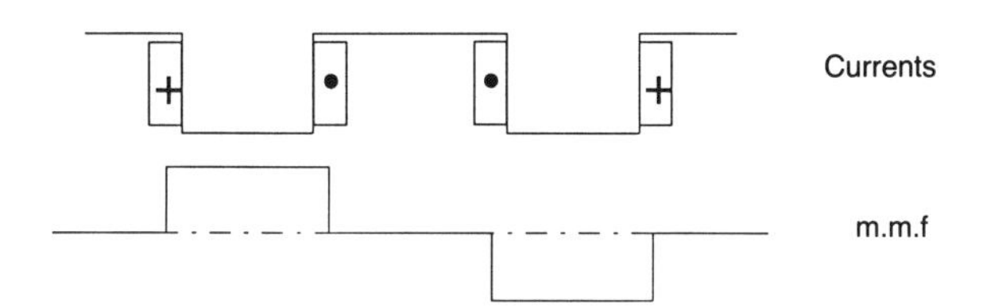

Fig. 4.12 m.m.f. diagram for a salient pole winding.

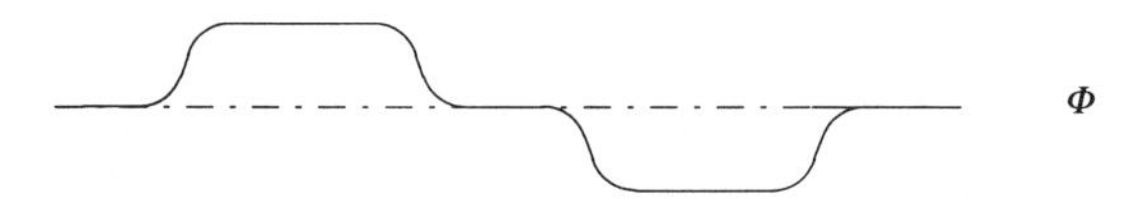

Fig. 4.13 Flux distribution associated with the m.m.f. diagram of Fig. 4.12.

The effective flux in the machine is then the combination of the field and armature fluxes. It must be noted that this is not the algebraic sum of the individual fluxes produced by the field and armature windings, as the **magnetic saturation** of the steel results in a net loss of flux. This effect of the armature field, and hence of armature current, in modifying the flux distribution in the rotating machine is referred to as **armature reaction** and will influence the performance of the machine under load.

The effects of armature reaction can be controlled by a number of techniques, all of which are intended to reduce the influence of armature current on the net flux in the machine. Techniques include:

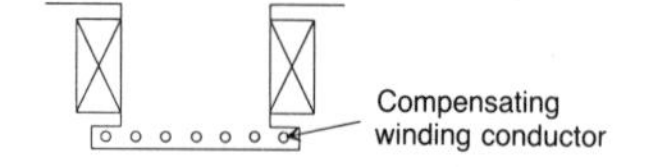

Fig. 4.14 Compensating winding buried in pole face of a salient pole.

- increasing the reluctance of the system by increasing pole tip saturation to control the effective reluctance profile of the airgap;
- the use of eccentric pole faces to control the reluctance profile of the airgap;
- compensating windings buried in the pole face and carrying armature current as in Fig. 4.14 to produce an m.m.f. in opposition to that of the armature itself thereby eliminating its contribution to the net field.

4.5 Commutation

As has been indicated, the current distribution in the armature of a d.c. machine is controlled by the commutator which acts to reverse the direction of current in the armature coils as they pass through the neutral axis between successive poles.

Consider the sequence of Fig. 4.15 in which, for simplicity, the brush is the same width as the commutator segment. With the brush positioned as in Fig. 4.15(a), current is transferred from the brush to coils I and II via the commutator segment as shown. After the commutator has rotated by an amount equivalent to one-quarter of the width of a commutator segment relative to the brush, the brush will have taken up the position in Fig. 4.15(b). Assuming an even distribution of current, the current will now be split between coils I, II and III in the manner shown. After a further rotation equivalent to a quarter of a commutator segment, the position in Fig. 4.15(c) will be reached in which no current flows in coil II. A further rotation of a quarter of a commutator segment will result in the position of Fig. 4.15(d) from which it is seen that the direction of the current in coil II has been reversed.

Comment

Commutation affects only those coils directly connected to the brushes via the appropriate commutator segments. The overall current distribution in the armature is unaffected.

A commutation process as described above would result in a reversal of current in coil II in the manner of Fig. 4.16(a) and is referred to as **straight line commutation**. In practice, a uniform transfer of current maintaining a constant current distribution at the brush would not occur as the self and mutual inductances associated with the armature coils result in the production of a voltage, referred to as the **reactance voltage** which opposes the change in current taking place. The result is a tendency for the

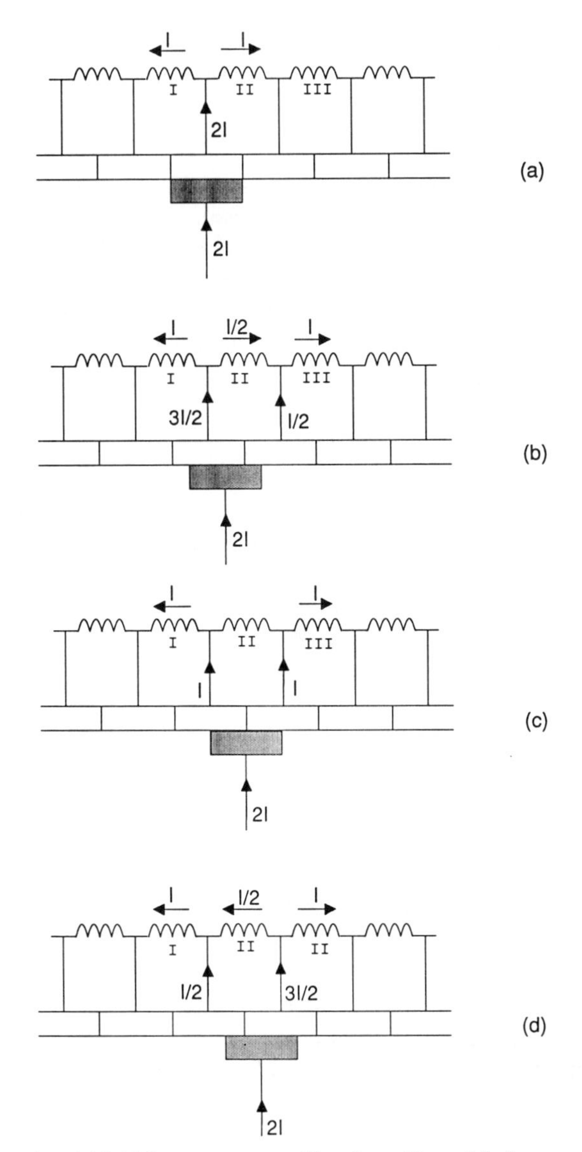

Fig. 4.15 Commutation: (a) initial commutator and brush positions; (b) after commutator movement equivalent to $\frac{1}{4}$ brush width; (c) after commutator movement equivalent to $\frac{1}{2}$ brush width; (d) after commutator movement equivalent to $\frac{3}{4}$ brush width.

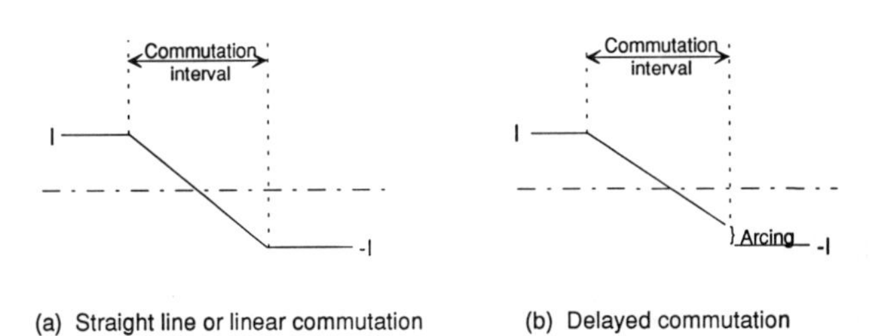

Fig. 4.16 Coil current during commutation: (a) straight line or linear commutation; and (b) delayed commutation.

current density to increase at the trailing edge of the brush resulting in **delayed commutation** or **undercommutation** with **trailing edge arcing** occurring as suggested by Fig. 4.16(b).

Commutation can be improved by:

- increasing brush contact resistance – this increases the brush volt drop, forcing a more uniform current distribution in the brush;
- the use of a multipolar design with a short armature and few turns;
- the use of interpoles. These as their name suggests are located between the main field poles as in Fig. 4.17. The interpoles carry armature current and produce an m.m.f. which induces a voltage in the armature coils in opposition to the reactance voltage, reducing its influence on the commutation process.

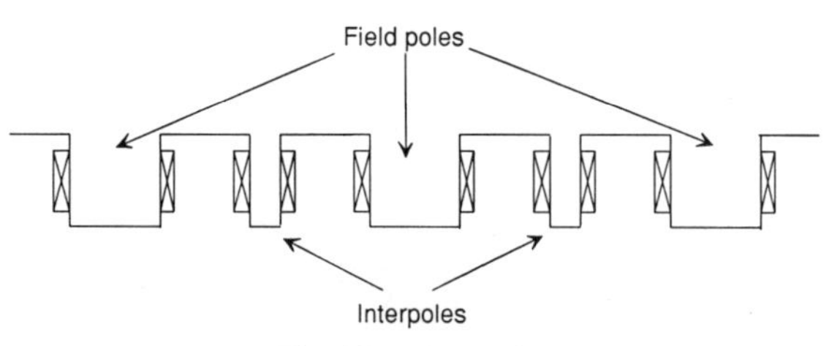

Fig. 4.17 Interpoles.

4.6 Field circuit

The field of a conventional d.c. machine is generated by the passage of the field current through the coils of the field winding and may use either **series field** winding, a **shunt field** winding or a **separately excited field** winding or some combination of these.

4.6.1 *Series field*

In a series connected d.c. machine all or part of the armature current flows through the field winding as in Fig. 4.18. The proportion of armature current in the field circuit may be controlled by the **diverter resistance** included in this figure.

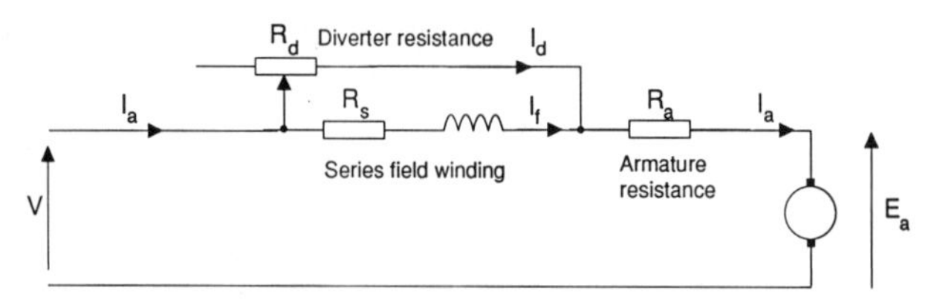

Fig. 4.18 d.c. machine with a series field and diverter resistance.

4.6.2 *Shunt field*

In a shunt connected d.c. machine the field current is derived from the same supply as the armature, as in Fig. 4.19. The field current is, however, independent of the armature current.

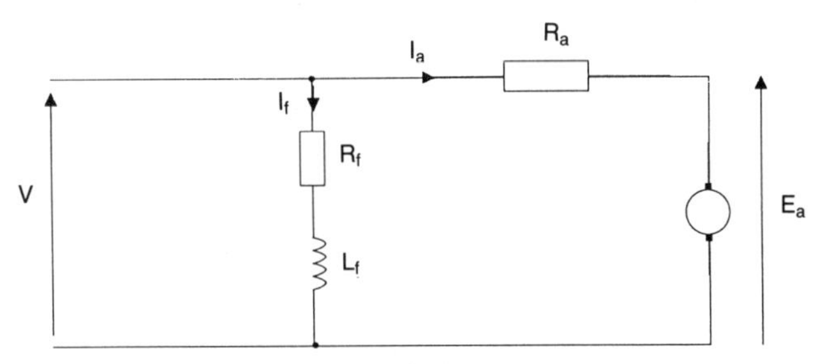

Fig. 4.19 d.c. machine with a shunt field.

4.6.3 *Separately excited field*

In a separately excited d.c. machine the field current is derived from its own independent supply as in Fig. 4.20.

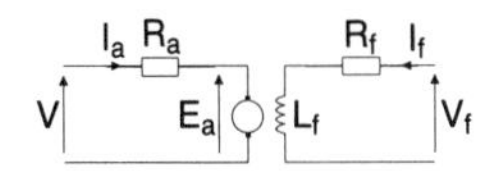

Fig. 4.20 d.c. machine with a separately excited field.

4.6.4 *Compound field*

A compound field winding contains both series and shunt (or separate) field windings as illustrated by Fig. 4.21. The resulting field can either be the sum – **cumulative compounding** – or the difference – **differential compounding** – of the applied series and shunt fields. In either case, the resultant field can be expressed in terms of an **equivalent shunt field current** I'_{field} obtained from the m.m.f. equation:

$$F_{\text{fld}} = N_{\text{shunt}} \cdot I_{\text{shunt}} \pm N_{\text{series}} \cdot I_{\text{series}} \tag{4.19}$$

when

$$I'_{\text{field}} = I_{\text{field}} \pm \frac{N_{\text{series}}}{N_{\text{shunt}}} \cdot I_{\text{series}} \tag{4.20}$$

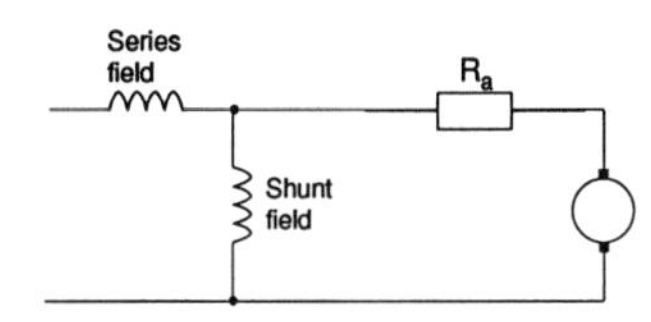

Fig. 4.21 d.c. machine with a compound field winding.

4.7 Open circuit characteristic or excitation curve

The **open circuit characteristic** or **excitation curve** of a d.c. machine has the general form of Fig. 4.22 and is obtained by driving the armature at constant speed and recording the open circuit armature voltage (E_a) for the full range of values of field current (I_f). The shape of the curve is determined by the magnetic characteristics the machine. Important features are:

- The production of a small voltage with zero applied field. This voltage results from the presence of the residual or remnant flux remaining in the ferro-magnetic material used to construct the machine once the main field has been removed.
- E_a is not a linear function of the field current I_f. This occurs because the machine is constructed from ferromagnetic materials which saturate at higher values of field, causing the rollover of the curve seen at the higher levels of field current.

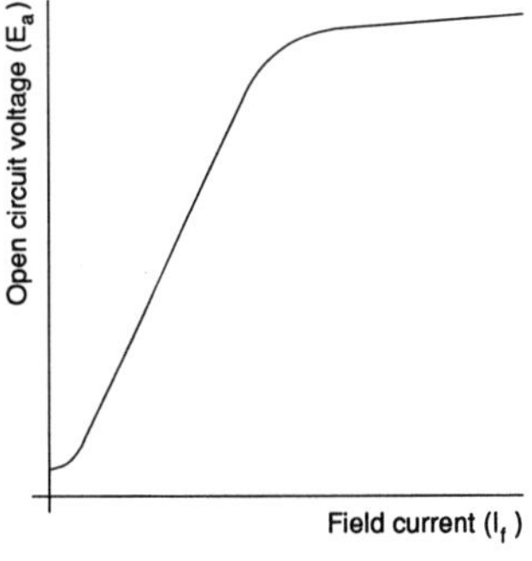

Fig. 4.22 Open circuit characteristic of a d.c. machine.

Only a single curve is needed to describe the operation of the d.c. machine over the whole of its speed range since for constant field the relationships:

$$\frac{E_{a0}}{E_{a1}} = \frac{\omega_{rm:0}}{\omega_{rm:1}}$$

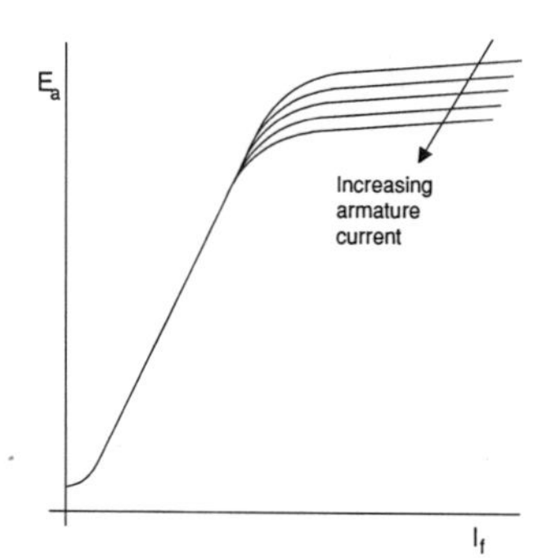

Fig. 4.23 Open circuit characteristic of a d.c. machine showing the effects of armature reaction.

and

$$E_{a1} = E_{a0} \cdot \frac{\omega_{rm:1}}{\omega_{rm:0}}$$

apply, enabling the back e.m.f. of the machine with a given value of field current to be obtained at any speed. By plotting E_a/ω_{rm} against I_f a general characteristic describing operation at any speed is obtained.

In practice, the effect of armature reaction will be to cause a net reduction in flux, and hence of E_a, at high values of field current leading to a modification of the open circuit characteristic as shown in Fig. 4.23.

4.8 Losses in d.c. machines

The losses within a d.c. machine can be broken down into **electrical losses**, **iron losses**, **mechanical losses** and **stray losses** as illustrated by Fig. 4.24 for the motoring condition. This situation would then be reversed when generating with the mechanical input power supplying the mechanical losses.

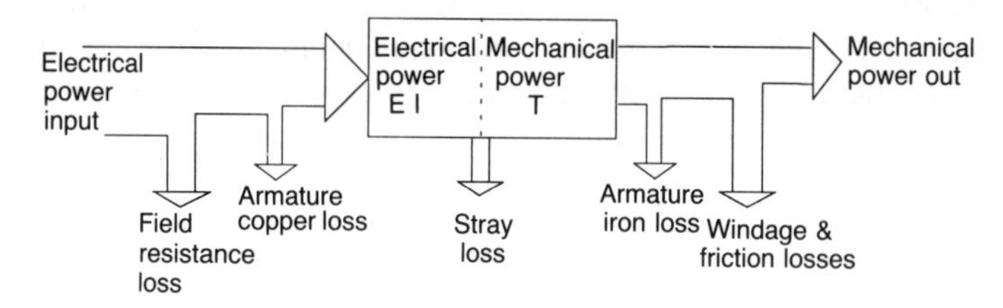

Fig. 4.24 Energy flow and losses in a d.c. motor.

4.8.1 *Electrical losses*

The principal electrical losses are the I^2R or copper losses in the field and armature circuits. In addition, there is a loss at the brushes which is a function of the current through the brush and the brush volt drop. For a typical carbon–graphite brush the effective brush resistance will decrease as the current through the brush increases with the result that the brush volt drop remains approximately constant over the operating range of the machine.

4.8.2 *Iron losses*

As the rotor of the d.c. machine is moving through the field produced by the field windings it will be subject to both hysteresis and eddy current losses. To reduce the eddy current loss in particular, the rotor of a d.c. machine will normally be assembled from laminations. There will also be some iron loss in the pole faces as a result of the flux ripples arising from the passage of the rotor teeth beneath the pole. In general however the poles will be manufactured from solid material except in the case of special machines such as servomotor machines which may be subject to rapid changes of field, in which case the poles will also be of a laminated construction.

As the iron losses are associated with the motion of the rotor they are a form of **rotational loss**.

4.8.3 *Mechanical losses*

The mechanical losses are a rotational loss associated with the motion of the rotor and may be expressed in terms of the friction effects (bearings, brush friction) and the air resistance effects (windage) experienced during rotation. Mechanical losses are usually grouped together as **windage and friction**.

4.8.4 *Stray loss*

The term **stray loss** is used to group together a number of effects such as **skin effect** in armature conductors and additional iron loss arising from leakage fluxes and is typically of the order of 1% of the rated power of the machine.

4.9 d.c. motors – operation

4.9.1 *Shunt and separately excited motors*

Consider the shunt and separately excited motors of Fig. 4.25. In Figs 4.25(a) and 4.25(b) the field current is controlled by means of the additional resistance R_f' incorporated into the field circuit while in Fig. 4.25(c) an independently variable voltage source is used to supply the field. Each of these connections is however identical in relation to the operation of the armature circuit.

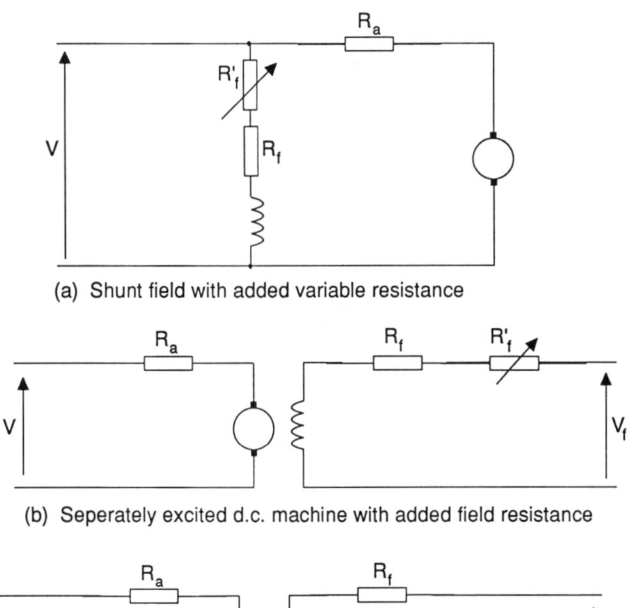

(a) Shunt field with added variable resistance

(b) Seperately excited d.c. machine with added field resistance

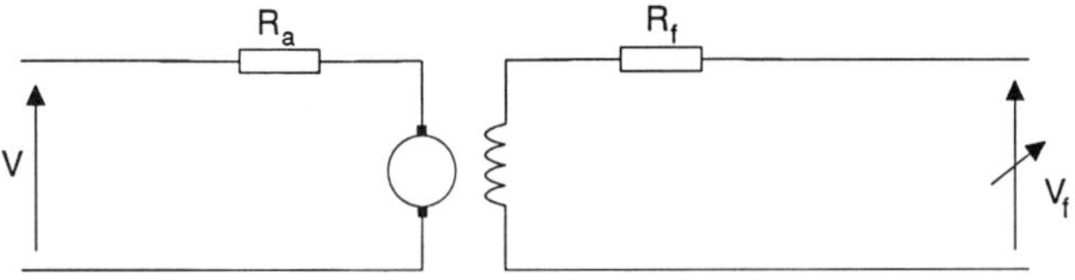

(c) Seperately excited d.c. machine with varaible field voltage

Fig. 4.25 d.c. shunt machine with field control: (a) shunt field with added variable resistance; (b) separately excited d.c. machine with added field resistance; and (c) separately excited d.c. machine with variable field voltage.

The torque/speed characteristics for shunt and separately excited machines are therefore identical and may be developed by reference to the torque and voltage equations as follows:

$$E_a = K\Phi\omega_{rm} \tag{4.21}$$

$$T = K\Phi I_a \tag{4.22}$$

Ignoring brush volt drop

$$E_a = K\Phi\omega_{rm} = V - I_a R_a \tag{4.23}$$

Hence

$$\omega = \frac{V}{K\Phi} - \frac{T.R_a}{(K\Phi)^2} \tag{4.24}$$

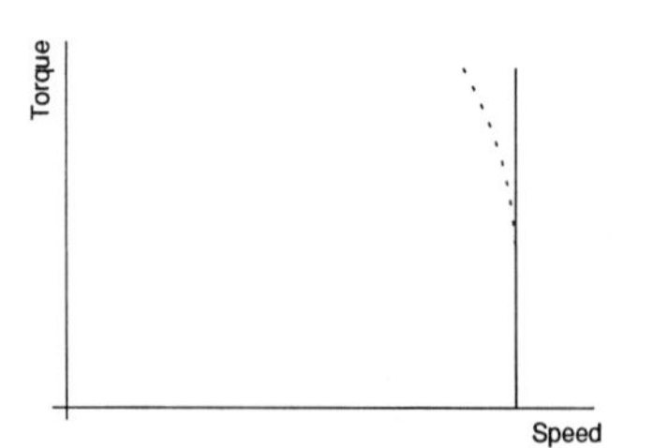

Fig. 4.26 Torque/speed characteristic of a d.c. shunt or separately excited motor.

Thus for constant field current, and hence constant Φ, the speed of the machine will vary linearly with torque. In practice the term $TR_a/(K\Phi)^2$ will be small in relation to the $V/K\Phi$ term and the speed variation will be of the order of a few per cent over the torque range as shown by the solid line in Fig. 4.26.

The effect of armature reaction in reducing the total available flux may become significant at higher torques resulting in a more rapid reduction in speed (**droop**) in this region as indicated by the dotted line in Fig. 4.26.

Example 4.1 Separately excited d.c. motor
A separately excited four-pole d.c. motor has an armature resistance of 0.72 ohms. When driving a particular load at 1400 rev/min it draws a current of 24 A from a 600 V d.c. supply. The mechanical losses (windage and friction) are 520 watt and are proportional to the speed of the machine. Find the load torque and the efficiency of the machine under these conditions. If the machine now provides the same load torque at a speed of 1275 rev/min, what will be the new values of field current, armature current and efficiency?

The open circuit characteristic of the machine at 1500 rev/min shows the following relationships:

I_f	0	0.2	0.4	0.6	0.8	1.0	1.2	1.4	1.6	A
E_a	50	115	215	322	426	520	600	627	637	V

At 1400 rev/min

$$E_a = 600 - 24 \times 0.72 = 582.72 \text{ V}$$

$$\omega_{rm} = \frac{2.\pi.1400}{60} = 146.61 \text{ radians/s}$$

Now

$$T_L = \frac{E_a.I_a - 520}{\omega_{rm}} = \frac{24 \times 582.72 - 520}{146.61} = 91.84 \text{ N m}$$

Power in $= 24 \times 600 = 14\,400$ watt
Power out $= 91.84 \times 146.61 = 13\,465$ watt

Therefore

$$\text{Efficiency} = \frac{13\ 465 \times 100}{14\ 400} = 93.5\%$$

At 1275 rev/min

$$\omega_{rm} = \frac{2\,.\,\pi\,.\,1275}{60} = 133.52 \text{ radians/s}$$

$$\text{Mechanical losses} = 520\left(\frac{1275}{1400}\right) = 473.6 \text{ watt}$$

Therefore

$$E_a\,.\,I_a = 91.84 \times 133.52 + 473.6 = 12\ 740 \text{ watt}$$

Substitute in armature voltage equation for I_a

$$600 = E_a + 0.72\left(\frac{12\ 740}{E_a}\right)$$

Solving

$$E_a = 584.3 \text{ V}$$

Therefore

$$I_a = 21.8 \text{ A}$$

$$\text{Efficiency} = \frac{91.84 \times 133.52 \times 100}{600 \times 21.8} = 93.8\%$$

Now

$$E_{a:1500} = E_{a:1275}\left(\frac{1500}{1275}\right) = 595.7 \text{ V}$$

Hence from open circuit characteristic

$$I_f = 1.19 \text{ A}$$

4.9.2 *Series motor*

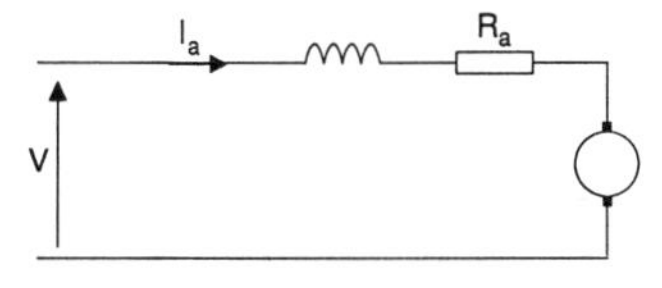

Fig. 4.27 d.c. series motor.

For the d.c. series motor of Fig. 4.27 the field and hence the back e.m.f. of the armature are functions of the armature current, resulting in a complex interrelationship between the torque, armature current and speed than is the case for the shunt or separately excited motors. The general shape of the torque/speed characteristic can, however, be obtained by using the approximate excitation characteristic of Fig. 4.28. In the lower region of this characteristic, flux and hence E_a is directly proportional to armature current, when, since:

$$\boldsymbol{\Phi} = \mathrm{K_s} I_a \tag{4.25}$$

then

$$T = (\mathrm{K}\,.\,\mathrm{K_s})\,.\,I_a^2 \tag{4.26}$$

in which the constant K is defined by the construction of the machine.

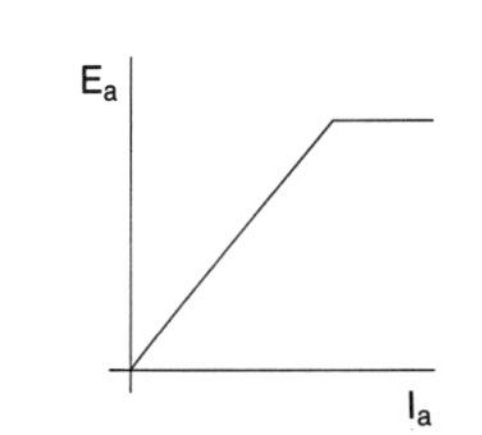

Fig. 4.28 Approximate open circuit characteristic for a d.c. machine.

Therefore

$$I_a = \left(\frac{T}{K . K_s}\right)^{\frac{1}{2}} \tag{4.27}$$

when, from the armature voltage equation:

$$\omega_{rm} = \frac{V}{K . K_s . I_a} - \frac{(R_a + R_s)}{K . K_s} = \frac{V}{(K . K_s)^{\frac{1}{2}} T^{\frac{1}{2}}} - \frac{(R_a + R_s)}{K . K_s} \tag{4.28}$$

The term $(R_a + R_s)/K . K_s$ is small and can generally be ignored when:

$$\omega_{rm}^2 = \frac{V^2}{K . K_s} . \frac{1}{T} \tag{4.29}$$

In the upper part of the characteristic of Fig. 4.28 flux is constant when:

$$E_a = K . K_s' . \omega_{rm} \tag{4.30}$$

in which the constant K_s' defines the effective flux level.

and

$$T = K . K_s' . I_a \tag{4.31}$$

This gives

$$\omega_{rm} = \frac{V}{K . K_s'} - \frac{(R_a + R_s) . T}{(K . K_s')^2} \tag{4.32}$$

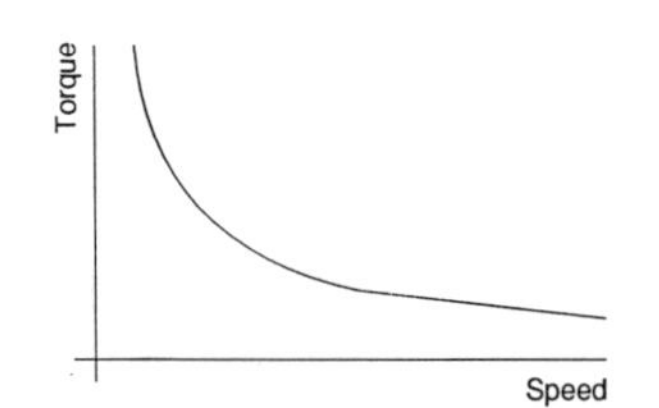

Fig. 4.29 Torque/speed characteristic of a d.c. series motor.

Combining equations (4.29) and (4.32) gives the overall torque/speed characteristic of Fig. 4.29 from which it can be seen that the d.c. series motor produces a high torque at low speeds and that this torque decreases rapidly with speed. A characteristic of this form is very well suited to traction loads such as railway locomotives, and d.c. series motors have been used extensively for this purpose.

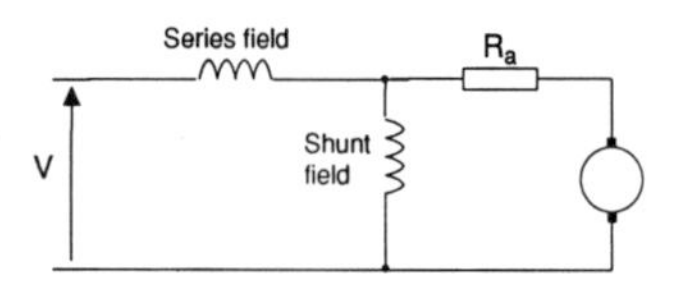

Fig. 4.30 d.c. compound motor.

Comment

Other advantages of series motors in traction applications are their cost, simplicity and reliability.

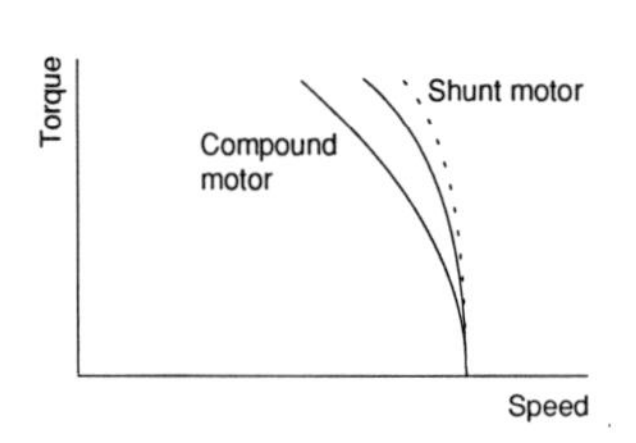

Fig. 4.31 Torque/speed characteristic of a cumulatively compounded d.c. motor.

4.9.3 *Compound motor*

The compound motor combines both series and shunt fields as in Fig. 4.30. Where compounding is used it is usually in the form of **cumulative compounding** (shunt and series fields add), reducing the armature current required for a particular value of torque and helping to offset armature reaction effects. The resulting torque/speed characteristic has the form shown in Fig. 4.31, together with that of a shunt motor for comparison.

Differential compounding (shunt and series fields subtract) results in a reduction of field with increasing armature current which could result in a speed instability in operation. The torque/speed characteristic for a differentially compounded machine is shown in Fig. 4.32, together with that of a shunt motor for comparison.

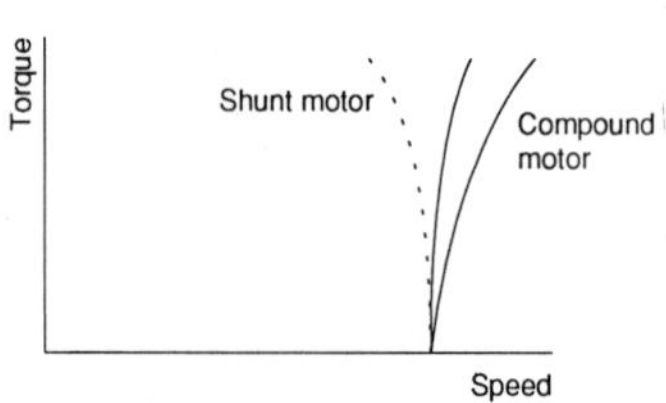

Fig. 4.32 Torque/speed characteristic of a differentially compounded d.c. motor.

Example 4.2 d.c. series motor
A d.c. series motor draws a current of 210 amperes from a 580 V supply when delivering its full load torque at a speed of 1200 rev/min. Given that the field and armature resistances are 0.14 ohms and 0.12 ohms, respectively, and that the load torques varies as the square of speed find the line current and steady state speed when the supply voltage is reduced to 420 V.

Saturation may be neglected.

Initially

$$E_{a1}=580-210(0.14+0.12)=525.4 \text{ V}$$

As saturation can be neglected operation corresponds to the linear portion of the curve of Fig. 4.27 when

$$T \propto I_a^2$$

Since

$$T \propto n^2$$

$$I_a^2 \propto n^2$$

and

$$\frac{I_{a1}}{I_{a2}}=\frac{n_1}{n_2}$$

Therefore

$$I_{a2}=0.175 \,.\, n_2 \qquad \text{(i)}$$

Also ignoring saturation

$$E_a \propto n \,.\, I_a$$

when

$$\frac{E_{a1}}{E_{a2}}=\frac{n_1 \,.\, I_{a1}}{n_2 \,.\, I_{a2}}$$

Thus

$$E_{a2}=2.085 \times 10^{-3} \,.\, n_2 \,.\, I_{a2} \qquad \text{(ii)}$$

Also

$$E_{a2}=420-0.26 I_{a2} \qquad \text{(iii)}$$

From (i), (ii) and (iii)

$$n_2^2+124.7 \,.\, n_2-1\ 151\ 080=0$$

Solving

$$n_2=1012.3 \text{ rev/min}$$

When from (i)

$$I_{a2}=177.2 \text{ A}$$

4.10 d.c. generators – operation

In a d.c. generator mechanical energy is supplied to the rotor via the shaft and is converted to electrical energy in the armature of the machine. It should be remembered that the input mechanical power provides not only the electrical power output but also the rotational losses (windage and friction, etc) of the machine. Hence:

$$\text{Mechanical power converted to electrical power } (P_{me}) = \text{Input mechanical power } (P_m) - \text{Rotational losses} \quad (4.33)$$

For any d.c. generator the output voltage is expressed in terms of the generated voltage as:

$$V = E_a - I_a R_a \quad (4.34)$$

and E_a is determined by the speed of the machine and the applied field by the relationship:

$$E_a = K\Phi\omega_{rm} \quad (4.35)$$

However, in the relationships:

$$T = K\Phi I_a \quad (4.36)$$

and

$$T.\omega_{rm} = E_a I_a \quad (4.37)$$

It must be remembered that the torque (T) is not the total mechanical torque applied at the shaft but that component of torque corresponding to P_{me} in equation (4.33).

4.10.1 *Separately excited generator*

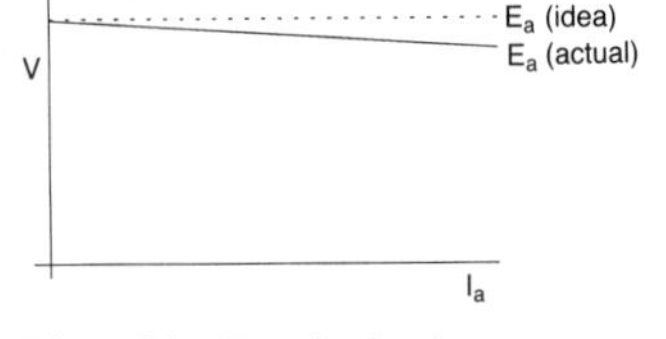

Fig. 4.34 Terminal voltage characteristic of a separately excited d.c. generator.

In the separately excited generator of Fig. 4.33 the generated voltage (E_a) is directly controlled by varying the field current when operating of constant speed. For variable speed operation, the field current may be adjusted to maintain either the generated voltage or the terminal voltage constant. For operation at constant field current and constant speed the terminal voltage will vary with armature current because of the $I_a R_a$ volt drop as shown in Fig. 4.34.

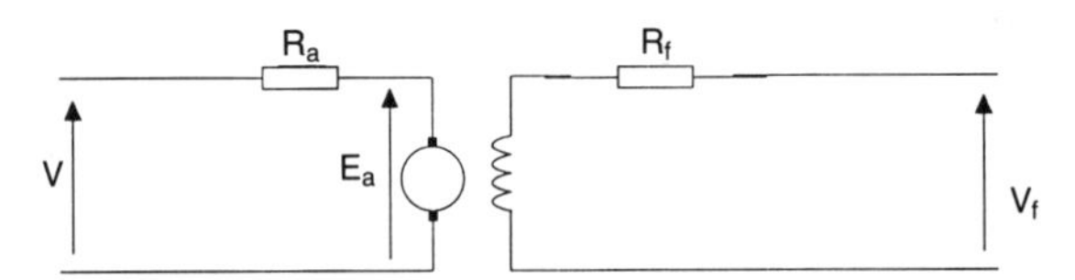

Fig. 4.33 Separately excited d.c. generator.

Example 4.3 Separately excited generator
A separately excited generator rated at 15 kW is used by an aircraft and is driven via gears from an auxiliary turbine; the speed variation from maximum to minimum is in the ratio 3:2. When operating at minimum speed on no load with a field current of 2.4 A the generator produces a voltage of 42 volts at its terminals.

Determine the maximum and minimum values of field current required if the terminal voltage of the generator is to be held constant at 42 volts under all conditions of load. The armature resistance of the generator is 0.0032 Ω and saturation effects may be neglected.

From information given and using $E_a = K' . I_f . \omega$ since saturation can be neglected, then:

$$K' . \omega_{min} = 17.5$$

in which case

$$K'\omega_{max} = 26.25$$

Now, minimum field is required at maximum speed on no-load in which case

$$I_{fmin} = \frac{42}{K'\omega_{max}} = 1.6 \text{ A}$$

Maximum field is required at minimum speed and full load when

$$I_a = \frac{15\,000}{42} = 357.1 \text{ A}$$

and

$$E_a = 42 + I_a . R_a = 43.14 \text{ V}$$

Hence

$$I_{fmax} = \frac{E_a}{K' . \omega_{min}} = 2.465 \text{ A}$$

4.10.2 *Shunt or self excited generator*

Consider the arrangement of Fig. 4.35 in which the field winding is connected directly across the terminals of the d.c. machine. If the machine is driven at constant speed with no load connected and switch 's' open there will be a small voltage at the terminals of the machine as a result of the residual magnetism in the magnetic circuit of the field. When switch 's' is closed a current will flow in the armature of the machine and through the field winding. Now, for this arrangement:

$$V = I_f R_f = -I_a R_a \tag{4.38}$$

and is shown as the field resistance line along with the excitation curve for the machine in Fig. 4.36(a). Provided the field produced by the current in the field winding is in the same direction as the residual field the result will be an increase in the generated voltage

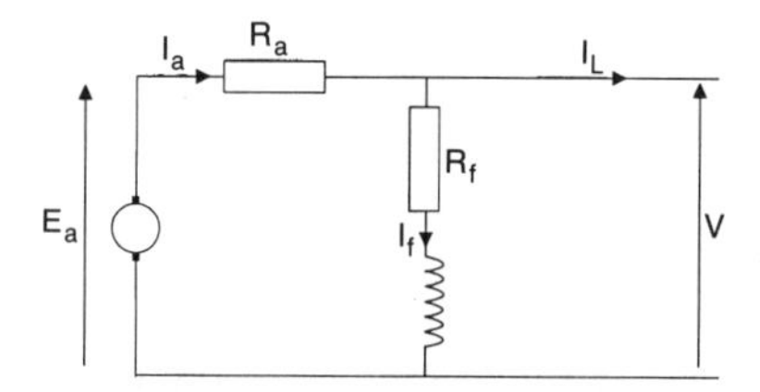

Fig. 4.35 Self excited d.c. generator.

and hence in the armature current. This increase in armature current is accompanied by a further increase in the field causing the generated voltage to increase. This buildup of generated voltage will continue until equilibrium is reached at which point the current in the field circuit matches the field requirements for the production of the driving voltage. For the case where the armature circuit resistance is small in comparison with the field resistance this occurs at the point at which the field resistance line intersects the open circuit characteristic for the machine – point 'a' on Fig. 4.36(a).

If load is now connected across the terminals of the machine:

$$V = I_L R_L = I_f R_f \quad (4.39)$$

$$I_a = I_L + I_f \quad (4.40)$$

and

$$E_a = V + I_a R_a \quad (4.41)$$

Referring to the open circuit characteristic of Fig. 4.36(b), these conditions are satisfied when the difference between the field resistance line and the open circuit characteristic matches the $I_a R_a$ value for a particular value of I_f – the condition shown as point 'b' on Fig. 4.36(b).

Referring to Fig. 4.37, if the value of R_f is too high then the field resistance line will intersect the open circuit characteristic at a low value of voltage and the machine will 'fail to self excite'. As R_f is reduced the slope of the field resistance line will reduce, eventually reaching a point at which it lies along the linear portion of the open circuit characteristic. This value of R_f is referred to as the **critical resistance** for the particular speed of operation and varies with the speed of the machine. If R_f is decreased below the critical resistance the machine will self excite.

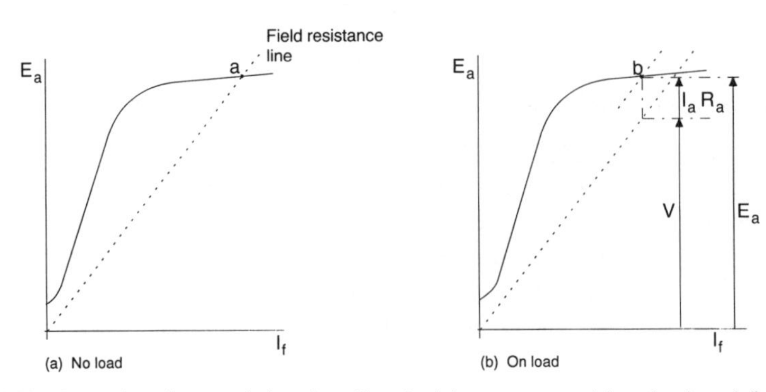

Fig. 4.36 Operating characteristics of a self excited d.c. generator: (a) no load; and (b) on load.

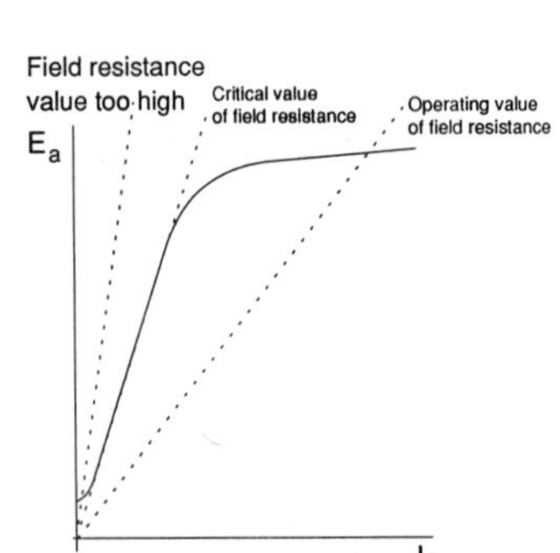

Fig. 4.37 Effect of field resistance variation on the operation of a self excited d.c. generator.

Example 4.4 Self-excited generator

The excitation characteristic of a self-excited generator running at 1100 rev/min is shown in Fig. 4.38. For a field resistance of 120 Ω and a speed of 1100 rev/min, what will be the open-circuit voltage of the generator?

If a resistive load drawing 350 A is now connected across the terminals of the machine what will be the new terminal voltage of the machine and the power in the load? The speed of the machine remains constant at 1100 rev/min throughout.

The armature circuit of the machine has a resistance of 0.068 Ω.

(a) From curve using field resistance line

$E_a = 256$ V

(b)

$I_a R_a = 350 \times 0.068 = 23.8$ V

Draw a line parallel to the field resistance line but displaced vertically by 23.8 V. The interception of this line with the open circuit characteristic gives the new operating point. Hence from Fig. 4.38:

Terminal voltage $= V = 225$ V

and

Load power $= 350 \times V = 78.75$ kW

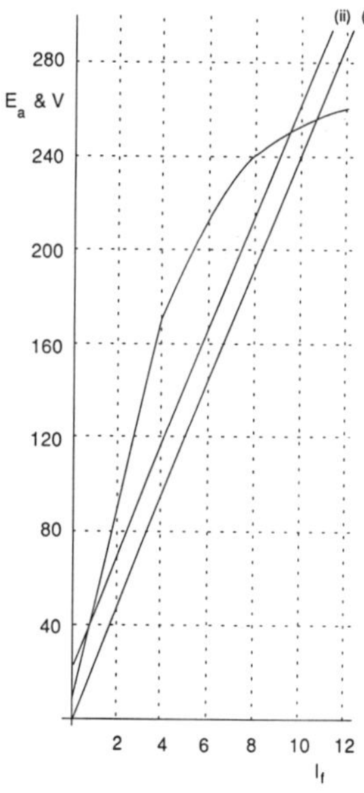

Fig. 4.38 Open circuit characteristic for Example 4.4.

4.11 d.c. machine braking

In variable speed applications a d.c. motor will be required to both accelerate and decelerate. Deceleration performance can be assisted by the use of one or other of the forms of electrical braking shown in Fig. 4.39.

4.11.1 *Resistive or dynamic braking*

In Fig. 4.39(a) the armature of the machine has been disconnected from the supply and connected to a resistor. In the presence of a field, the back e.m.f. (E_a) produced will dirve a current through this resistor, dissipating the energy stored in the rotating inertia of the machine. The braking effect achieved is a function of speed.

4.11.2 *Regenerative braking*

In the situation of Fig. 4.39(b) the field of the machine is adjusted so as to make $E_a > V$ when the machine will act as a generator returning energy to the supply and bringing the machine to rest. As speed decreases the field must be increased in order to maintain $E_a > V$. The maximum field is limited by the field voltage supply and the maximum permitted field current and this will set a limit on the speed range over which regenerative braking can be used.

4.11.3 *Reverse current braking or plugging*

By reversing the connections to the supply as in Fig. 4.39(c) the direction of current through the armature of the machine is reversed producing a reverse torque which rapidly decelerates the motor. This is a very severe condition and a current limiting resistor is normally included in the armature circuit to limit the current.

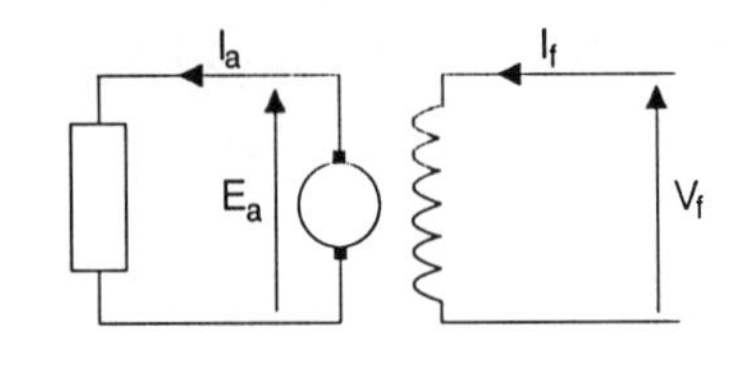

(a) Dynamic or resistive braking

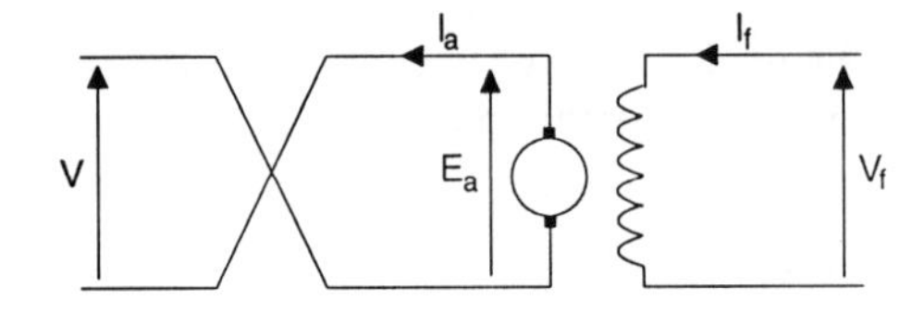

(b) Reverse current braking or plugging

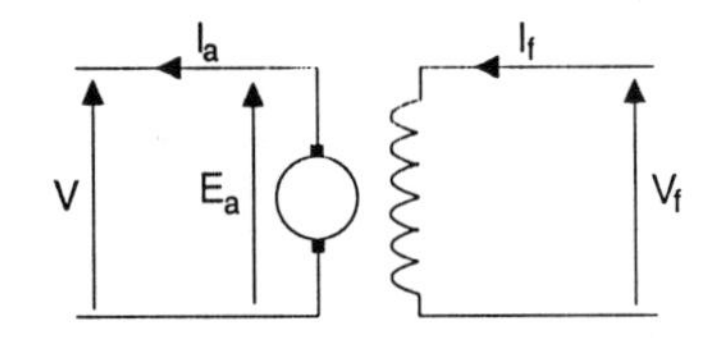

(c) Regenerative braking; $E_a > V$

Fig. 4.39 d.c. machine braking modes: (a) dynamic or resistive braking; (b) reverse current braking or plugging; and (c) regenerative braking $E > V$.

4.12 Speed control

4.12.1 *Field control*

If armature resistance is ignored then the equation of the armature circuit reduces to:

$$E_a = V = \mathrm{K}\boldsymbol{\Phi}\omega_{rm} \tag{4.42}$$

With the supply voltage (V) constant the relationship between the field current and the speed of the machine can be written as:

$$\omega_{rm} \propto \frac{1}{\boldsymbol{\Phi}}$$

Thus reducing the field $\boldsymbol{\Phi}$ of the d.c. machine by reducing the field current is associated with an increase in speed. Taking the rated value of armature current as the limiting condition the maximum torque delivered will be:

$$T_{max} = \mathrm{K_t} \,.\, \boldsymbol{\Phi} = \frac{\mathrm{K}'_t}{\omega_{rm}} \tag{4.43}$$

and the mechanical power developed under these conditions is:

$$P_m = T\omega_{rm} = \mathrm{K_t} \,.\, \mathrm{K}'_t \tag{4.44}$$

The controlled machine will therefore operate with a constant power boundary above the base speed ($\omega_{rm:base}$), as in Fig. 4.40.

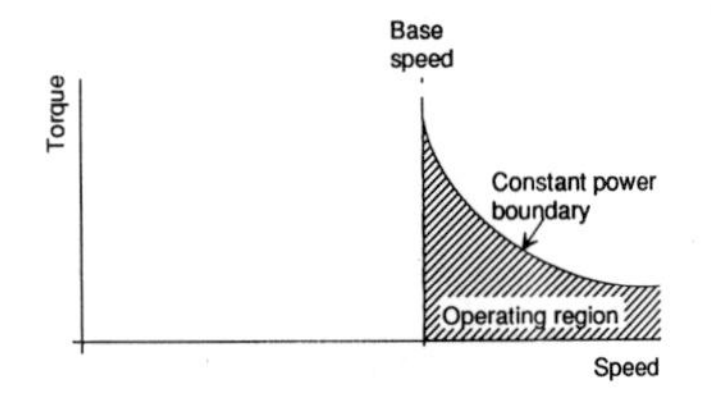

Fig. 4.40 Torque/speed characteristic of a d.c. machine with field control.

Comment

Base speed is defined by the application of maximum field current together with maximum applied armature voltage.

The reduction in field is usually limited to about 70% of the maximum field because of the increasing influence of armature reaction which may result in instability. Performance at low fields can be improved by the inclusion of a compensating winding (Fig. 4.14) or the use of a small amount of cumulative compounding.

The simplest means of providing field control is by the inclusion of series resistance in the case of a shunt or separately excited motor or through the use of a diverter resistance in the case of a series motor. In all cases protection against a complete loss of field must be incorporated to guard against the possibility of severe overspeeding of the machine with consequent risk to the user and damage to the machine.

4.12.2 *Armature voltage control*

Referring again to equation (4.42), if the field of the machine is maintained constant and the armature voltage varied then the relationship between this voltage and the speed of the machine may be approximated by:

$$\omega_{rm} \propto V$$

Taking the rated armature current as the limit as before the maximum torque is:

$$T_{max} = K_m \cdot I_a = \text{Constant} \quad (4.45)$$

and the machine operates with a constant torque boundary below the base speed as in Fig. 4.41.

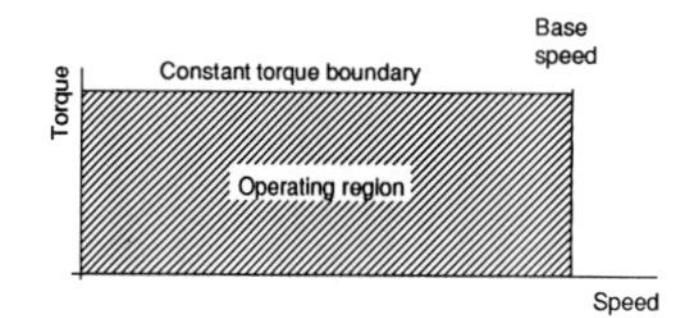

Fig. 4.41 Torque/speed characteristic of a d.c. machine with armature voltage control.

For a long time the most effective means of achieving armature voltage control was the **Ward Leonard** set of Fig. 4.42. This used a constant speed **prime mover** (induction machine, diesel engine, steam turbine) to drive a d.c. generator, the output voltage of which was controlled by varying the field. The output of the gnerator was then directly connected to the d.c. motor.

(a) *Thyristor control*

With the advent of the thyristor, the variable voltage supply could be provided directly from the a.c. system by means of either a fully controlled or a half-controlled converter or from a d.c. source by means of a controlled chopper circuit.

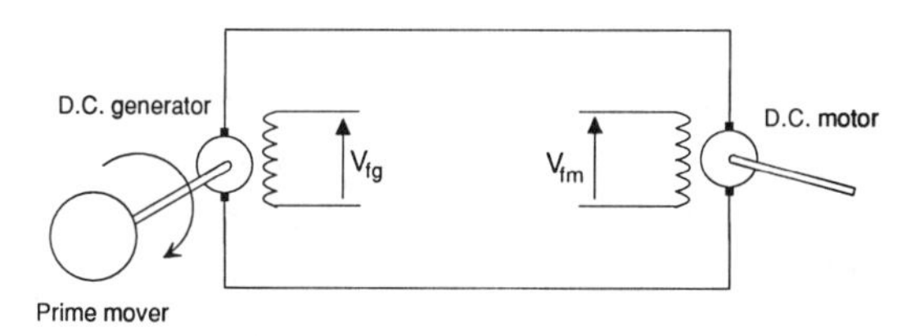

Fig. 4.42 Ward Leonard set.

Comment

See Chapter 3 for a discussion of fully controlled converters, half-controlled converters and d.c. choppers.

The basic configuration of a converter drive is shown in Fig. 4.43 using a fully controlled, six-pulse bridge converter. The majority of converter drives are intended for operation with a d.c. machine having sufficient armature inductance to maintain the armature current sensibly constant over one cycle of the fundamental waveform. For applications such as traction drives where the performance of the converter must be matched to that of the motor this necessary inductance would be a design feature of the armature circuit either integral to the motor itself or as a separate series inductance.

For lower rated drives using half-controlled converters with relatively low levels of armature circuit inductance, a freewheeling or bypass diode is included as in Fig. 4.44. This prevents the reversal of voltage at the terminals of the converter and results in improved commutation and a reduction of the armature current ripple. Under conditions of low load or on starting the armature current may, however, become discontinuous.

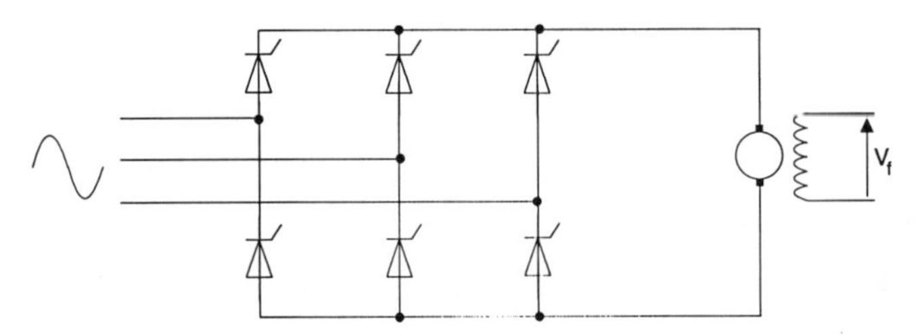

Fig. 4.43 Armature voltage control using a fully controlled bridge converter.

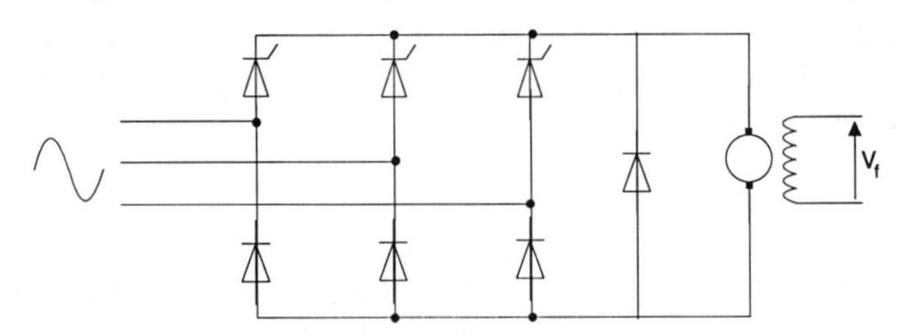

Fig. 4.44 Armature voltage control using a half-controlled bridge converter.

Resistive or dynamic braking can be applied by disconnecting the converter an connecting a resistor across the armature as in Fig. 4.39(a). The braking performanc can then be controlled to a certain degree by varying the field current of the machine

Two-quadrant operation involving regeneration can be achieved by means of a full controlled bridge and field reversal as suggested by Fig. 4.45. Care must be taken t ensure that the armature current is maintained within limits during the changeove period.

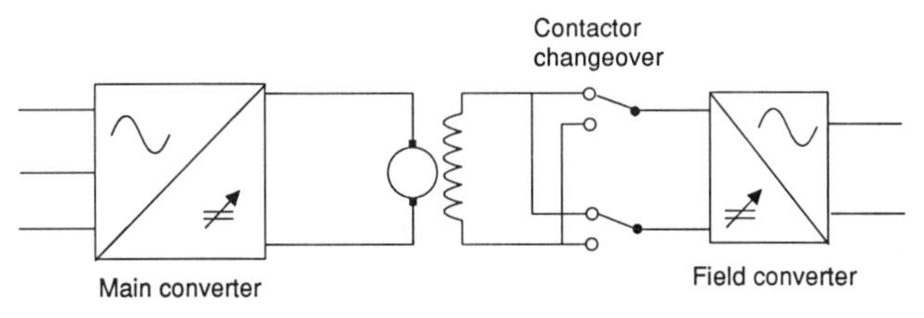

Fig. 4.45 Field reversal using contactor changeover.

Four-quadrant operation can be achieved by using a pair of bridges as in Fig. 4.46(a). Assuming initial operation in motoring mode using bridge B_1 then reversal would take place as follows.

1. The firing angle of bridge B_1 is increased to turn off the current in the bridge.
2. Bridge B_1 is turned off.
3. Bridge B_2 takes over in the inverting mode and the machine regenerates returning energy to the supply.
4. As the speed of the machine reduces the firing angle of bridge B_2 is reduced to give zero voltage at or near to zero speed.
5. The firing angle of bridge B_2 is further reduced to control the acceleration of the machine in the reverse direction.
6. The machine now runs as a motor in the reverse direction from bridge B_2.

An alternative solution combining field reversal with a motoring and a regenerating bridge is shown in Fig. 4.46(b). Here, load is transferred from bridge B_1 to bridge B_2 as before but as the speed of the machine approaches zero the field current is reversed using the field bridges F_1 and F_2 and the machine accelerates in the reverse direction using bridge B_1.

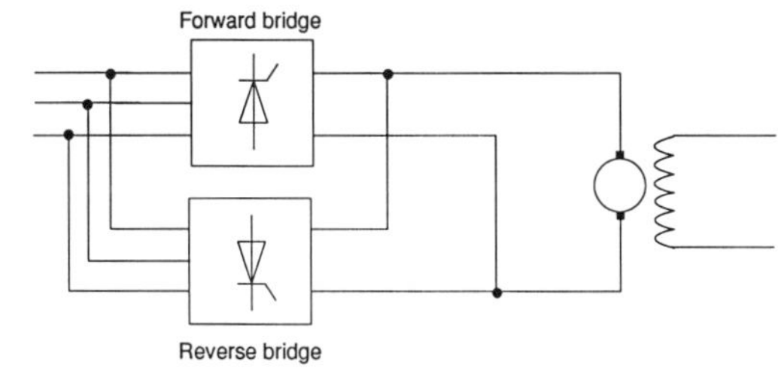

(a) 4-quadrant operation using a forward and a reverse bridge

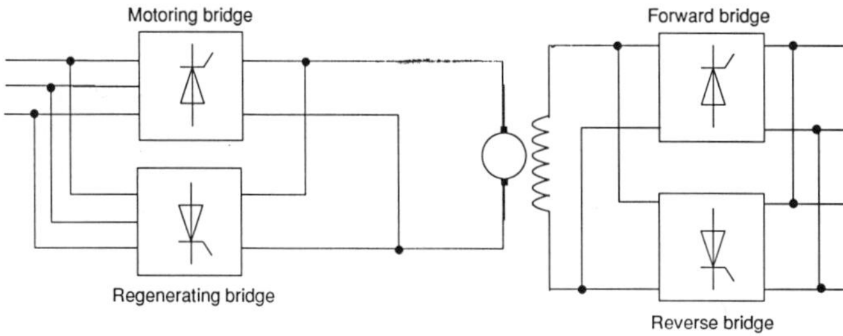

(b) 4-quadrant drive using motoring and generating bridges and forward and reverse field bridges

Fig. 4.46 Four-quadrant d.c. machine drives: (a) four-quadrant operation using a forward and a reverse bridge; and (b) four-quadrant drive using motoring and generating bridges and forward and reverse field bridges.

(b) *Chopper drives*

Figure 4.47 shows a separately excited d.c. machine being operated from a chopper circuit. Ignoring the small pulsations in machine torque, the speed of the machine may by assumed constant under steady state conditions when:

$$E_a = K_1 . \omega_{rm} \tag{4.46}$$

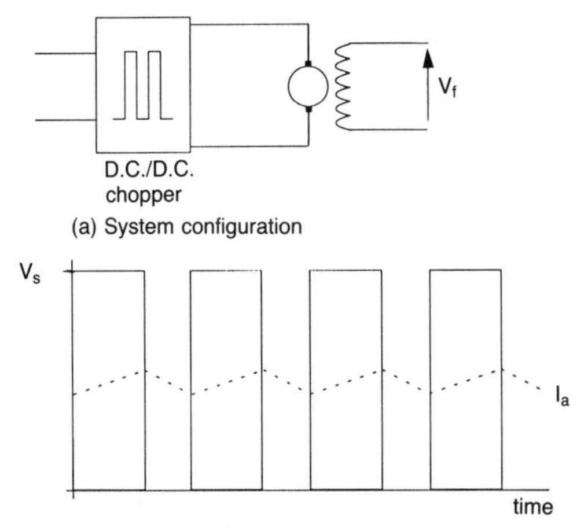

Fig. 4.47 d.c. machine with chopper drive: (a) system configuration; and (b) armature current with period much less than system time constant.

Comment

The inertia of the machine acts to smooth out the torque pulsations and maintain the speed.

Assuming a continuous armature current then during turn-on:

$$V_s = E_a + i_a . R_a + L_a . \frac{di_a}{dt} \tag{4.47}$$

Given that the period (T) of the chopped waveform is short in comparison with the system time constant the variation in armature current may be considered to be linear. Also, if the variation in armature current is small with respect to the mean value of armature current then the value of the $i_a . R_a$ term in equation (4.47), which is small in relation to V_s, may be considered to be sensibly constant when:

$$L_a . \Delta I_a = [V_s - (E_a + I_a R_a)] . t_1 \tag{4.48}$$

During turn-off

$$0 = E_a + i_a . R_a + L_a . \frac{di_a}{dt} \tag{4.49}$$

when

$$L_a \,.\, \Delta I_a = (E_a + I_a R_a) \,.\, t_2 = (E_a + I_a \,.\, R_a) \,.\, (T - t_1) \tag{4.50}$$

Combining equations (4.48) and (4.50) gives:

$$V_s = \left(\frac{T}{t_1}\right)(E_a + I_a \,.\, R_a) \tag{4.51}$$

In which case, using equation (4.46)

$$\omega_{rm} = \frac{V_s\left(\frac{t_1}{T}\right) - I_a \,.\, R_a}{K_1} \tag{4.52}$$

Torque is a function of the mean armature current such that

$$T = K_1 \,.\, I_a \tag{4.53}$$

and the input power is given by

$$P_{in} = V_s \,.\, I_a \,.\left(\frac{t_1}{T}\right) \tag{4.54}$$

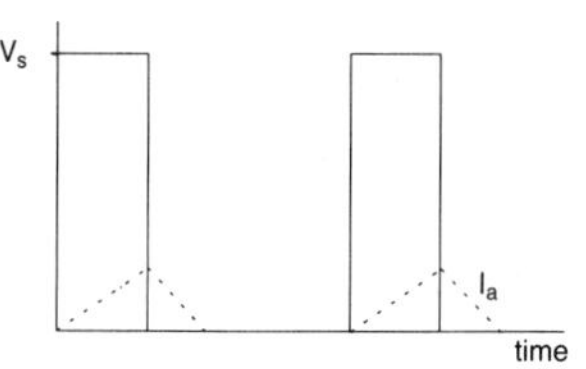

Fig. 4.48 Chopper drive with discontinuous current.

Under conditions of light load it is possible that the armature current of the machine may become discontinuous as shown in Fig. 4.48. In this case the $i_a R_a$ drop may be negleced in relation to the supply voltage when:

$$\Delta I_a = \frac{(V_s - E_a)t_1}{L_a} = \frac{E_a \,.\, t_2}{L_a} \tag{4.55}$$

giving

$$E_a = V_s \,.\left(\frac{t_1}{t_1 + t_2}\right) \tag{4.56}$$

in which case

$$\omega_{rm} = \frac{V_s \,.\left(\frac{t_1}{t_1 + t_2}\right)}{K_1} \tag{4.57}$$

The chopper may also be used to control the starting performance of a d.c. machine. Typically, during starting the chopper controller would vary the mark/space ratio of the pulses applied to the motor in order to keep the armature current within preset limits.

Example 4.5 Chopper drive

Consider a circuit of the form of Fig. 4.47 used to control the speed of a d.c. machine from a 620 V d.c. source. The armature parameter values of the machine are $L_a = 5.2$ mH and $R_a = 0.096\ \Omega$ and it has a field constant of 4.8 V/(radian/s). If the chopping frequency is 1 kHz, estimate the variation in load current when the machine is operating at 400 rev/min with a mean armature current of 110 A.

For the same mark/space ratio, what will be the minimum load current condition if continuous conduction is to be maintained? What will be the speed of the machine under these conditions?

$$E_a = \frac{4.8 \times 400 \times 2\pi}{60} = 201.1 \text{ V}$$

hence

$$V = 201.1 + I_a R_a = 211.66 \text{ V}$$

and

$$t_1 = \frac{211.66}{620} = 0.3414 \; ms$$

During on period

$$\Delta I_a = \frac{[V_s - (E_a + I_a R_a)]t_1}{L_a} = 26.81 \text{ A}$$

On minimum load with continuous current the mean value of I_a must be $\Delta I_a/2 = 13.4$ A, in which case

$$E_a = 211.66 - 13.4 \times 0.096 = 210.37 \text{ V}$$

Therefore

$$\text{Speed} = \frac{210.37}{4.8} = 43.83 \text{ radians/s} = 418.5 \text{ rev/min.}$$

Figure 4.47 shows single quadrant or **class A** chopper drive allowing energy to flow only from the d.c. supply to the motor. By adding a second switching element as in Fig. 4.49 a two-quadrant or **class C** drive is obtained. In order to regenerate, the motoring chopper C_1 is turned off when a decaying current circulates in the machine armature via freewheeling diode D_1. When this current reaches zero, the regenerating chopper C_2 is turned on. With C_2 on, a reverse current is driven through the armature circuit of the machine and C_2 by the back e.m.f. of the machine. When C_2 is turned off, proportion of the energy stored in the inductance of the armature circuit is returned to the supply via the regenerating diode D_2.

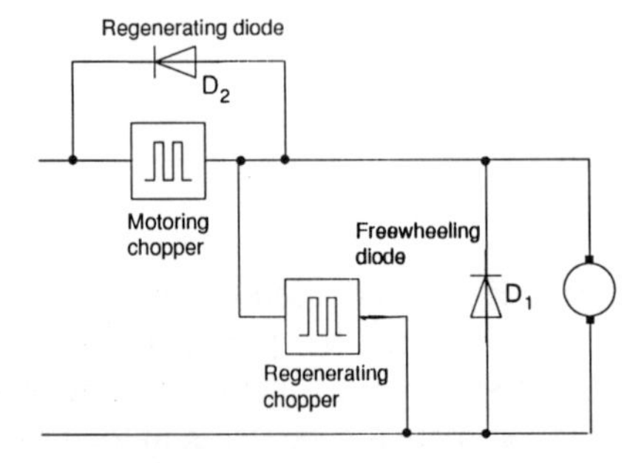

Fig. 4.49 Two-quadrant or class C chopper.

Comment

The control circuit must ensure that C_1 and C_2 are not on together as this would result in the short circuiting of the supply.

Full four-quadrant operation can be achieved with the circuit of Fig. 4.49 by reversing the field current of the motor at or near to zero speed. Alternatively, the **H-bridge** circuit of Fig. 4.50 can be used, removing the need for field reversal. The circuit of Fig. 4.50 can be operated in either **bipolar mode** with two transistors switching simultaneously or in **unipolar mode** with one transistor continuously on with a second providing the switching. Figure 4.51 illustrates the operation of the bridge when motoring in both bipolar and unipolar modes.

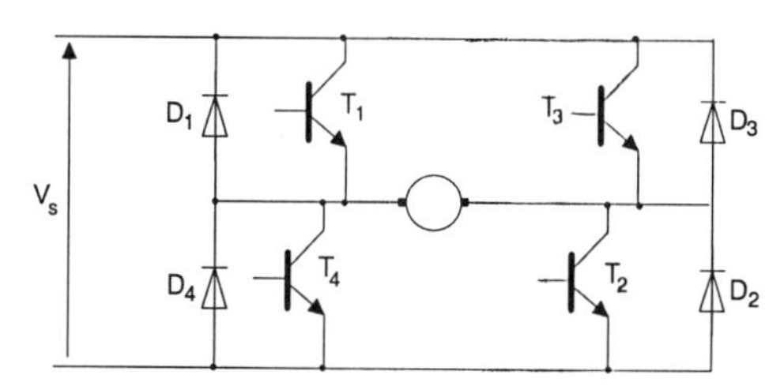

Fig. 4.50 H-bridge circuit.

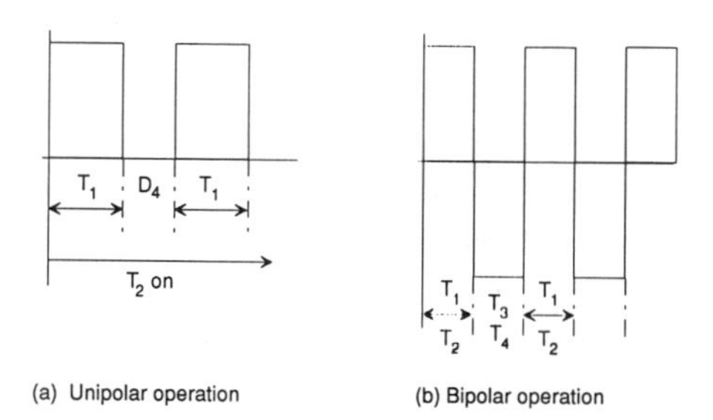

Fig. 4.51 Operation of H-bridge in motoring mode: (a) unipolar operation; and (b) bipolar operation.

4.13 d.c. machine dynamics

The machine equations can be rewritten as:

Field $$v_f = i_f \,.\, R_f + L_f \frac{di_f}{dt} \tag{4.58}$$

Armature $$v = e_a + i_a \,.\, R_a + L_a \frac{di_a}{dt} = K \,.\, f(i_f) \,.\, \omega_{rm} + i_a R_a + L_a \frac{di_a}{dt} \tag{4.59}$$

Torque $$T = K \,.\, f(i_f) \,.\, i_a = J \frac{d\omega_{rm}}{dt} + T_{mech} \tag{4.60}$$

in which T_{mech} is the mechanical load torque.

For a separately excited motor with constant field and armature voltage control.

$$T = K_m \cdot i_a \tag{4.61}$$

$$e_a = K_m \cdot \omega_{rm} \tag{4.62}$$

and

$$e_a = v - i_a R_a - L_a \frac{di_a}{dt} = v - i_a R_a - L_a \cdot s i_a \tag{4.63}$$

Hence

$$\frac{v - e_a}{R_a} = i_a \cdot (l + \tau_a s) \tag{4.64}$$

where τ_a is the armature time constant and s represents d/dt.

Also

$$J s \omega_{rm} + T_{mech} = T \tag{4.65}$$

Therefore

$$s\omega_{rm} = \frac{K_m i_a}{J} - \frac{T_{mech}}{J} \tag{4.66}$$

Example 4.6 d.c. machine dynamics: resistive starting
To prevent excessive current on starting a d.c. machine, resistance is introduced into the armature circuit which is then progressively switched out as the speed of the machine increases and the back e.m.f. builds up. Figure 4.52 shows the armature circuit of a d.c. shunt motor with a stepped resistive starter. Determine expressions for the variation of armature current and speed during run-up. Assume the upper limit of permitted current in the armature to be I_{a1} with switching occurring at an armature current of I_{a2}.

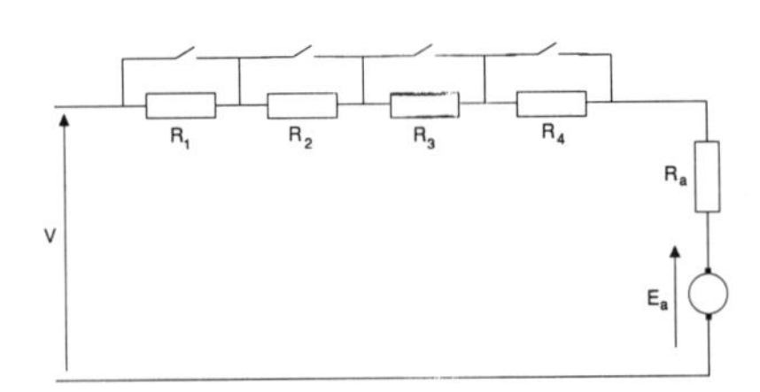

Fig. 4.52 Resistance starter for a d.c. machine.

With the d.c. machine at standstill no back e.m.f. is being produced and, if armature inductance effects are ignored, armature current will be limited only by the total resistance of the armature circuit. Referring to Fig. 4.52:

Hence, initially

$$E_a = 0$$

Setting the upper limit on armature current at I_{a1} and substituting in the expression

$$V = E_a + I_a . R_a$$

for the initial conditions of Fig. 4.52 gives:

$$V = I_{a1} R_{a1} = I_{a1} . (R_a + R_1 + R_2 + R_3 + \dots) \tag{i}$$

from which the total value of armature resistance required $(R_a + R_1 + R_2 + R_3 + \dots)$ can be found. After a time interval t_1, the back e.m.f. has increased to a value of E_{a1} and the armature current has fallen to I_{a2} when:

$$V = E_{a1} + I_{a2} . R_{a1} = E_{a1} + I_{a2} . (R_a + R_1 + R_2 + R_3 + \dots)$$

enabling E_{a1} to be found. If resistance R_1 is now switched out, such that the armature current increases to I_{a1}, when:

$$V = E_{a1} + I_{a1} . (R_a + R_2 + R_3 + \dots) \tag{ii}$$

Combining equations (i) and (ii) gives the value of R_1. This procedure is then repeated until all the added resistance in the armature circuit has been removed.

The form of the curves for speed, armature current and torque during starting can then be found from the dynamic equations as follows. Torque:

$$T = J \frac{d\omega_{rm}}{dt} + T_{mech} = Js\omega_{rm} + T_{mech}$$

Ignoring armature inductance L_a

$$V = e_a + i_a . R_a$$

With constant field:

$$e_a = K_m \omega_{rm}$$

when

$$i_a = \frac{V - K_m \omega_{rm}}{R_a}$$

Then

$$K_m . i_a = Js\omega_{rm} + T_{mech}$$

when

$$\frac{K_m V}{R_a} - \frac{K_m^2 . \omega_{rm}}{R_a} = Js\omega_{rm} + T_{mech}$$

Hence

$$\omega_{rm}\left(s + \frac{K_m^2}{J . R_a}\right) = \frac{K_m V}{J . R_a} - \frac{T_{mech}}{J}$$

The variation of ω_{rm} and I_a during run up is illustrated by Figs 4.53(a) and 4.53(b).

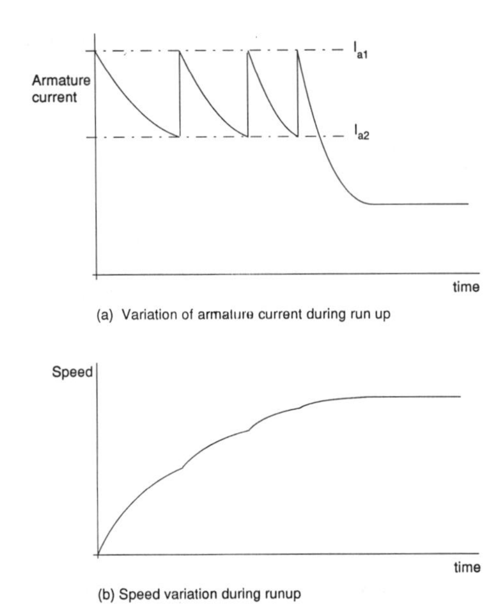

Fig. 4.53 Current and speed profiles of a d.c. machine during starting with a four-step starter.

4.14 Brushless d.c. machines

In a conventional d.c. machine the necessary field structure is provided by the field winding located on the stator of the machine. In a number of applications such as servodrives, the presence of the field winding offers no particular advantage and indeed in some circumstances may be a disadvantage, for instance because of the maintenance requirements of the brushgear.

In a brushless d.c. machine the field circuit is replaced by a powerful permanent magnet, eliminating the need for a field winding. The construction of the brushless machine typically places the field magnet on the rotor as in Figs 4.54(a) and 4.54(b), eliminating the need for a conventional commutator and associated brushgear. The stator winding now forms the armature of the machine and electronic commutation of the armature currents is used to provide the required relationship with the field.

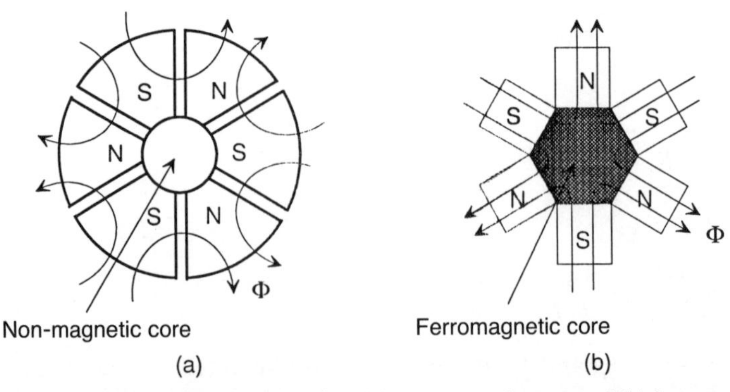

Fig. 4.54 Brushless machine, rotor construction: (a) non-magnetic core; and (b) ferromagnetic core.

Comment

Brushless machines are also referred to as permanent magnet (PM) motors. Alnico, ferrite (ceramic) and rare earth magnets such as samarium–cobalt and neodymium–iron–boron are used as the field magnet.

Brushless machines can either be operated as synchronous motors (Chapter 5) receiving a variable frequency multi-phase supply, or as d.c. motors with the commutation of the armature currents determined appropriately.

When operated as a d.c. machine the supply to the armature windings of Fig. 4.55 are switched in the sequence AB–AC–BC–BA–CA–CB–AB–etc. with switching taking place every 60° (electrical). The actual point at which the switching occurs is determined by reference to rotor position, either by directly monitoring the shaft position or by using Hall effect devices to sense the position of the magnetic field. This is referred to as **electronic commutation**.

Brushless machines can exhibit a ripple in their output torque. The major components of this ripple are the **reluctance ripple** resulting from the inherent magnetic asymmetry of the machine, a **drive current ripple** at a frequency of $pn/2$, where p is the number of poles and n is the speed in rev/min, and the **once round ripple**, at a frequency corresponding to the speed of rotation of the machine, caused by the error in alignment of the rotor within the stator.

Brushless machines have a number of advantages over conventional d.c. machines including:

- reduced maintenance;
- higher torque/volume ratios;
- improved torque characteristics, particularly at higher speeds;
- simplified protection,

and, as a result, they have found increasing application as servomotors in robots and machine tools.

The major disadvantages with brushless machines are those associated with the presence of a permanent magnet. A particular concern is the need to avoid any

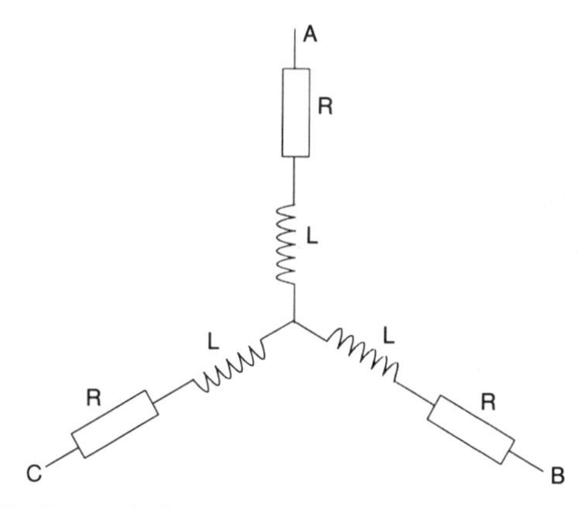

Fig. 4.55 Stator windings and connections for a brushless machine.

demagnetization of the magnet, for instance as a result of excessive armature currents or disassembly of the machine. Any such demagnetization would result in a reduction of the available field and a decrease in performance of the motor. Care also needs to be taken in the design of the motor casing to prevent the permanent magnet attracting to itself foreign particles which could inhibit operation and possibly damage the motor.

Exercises

1 When a 5.6 kW, 240 volt, 28.5 ampere, 960 rev/min d.c. shunt motor is driven at rated speed with a field current of 0.91 amperes an open circuit voltage of 240 volts is produced at the armature terminals.

The machine is now connected as a shunt motor to a 240 volt d.c. supply and the field current adjusted to 0.91 amperes, find:

(a) the speed at which the machine develops an internal torque equal to its rated torque;

(b) the speed at which this torque is developed if the field current is reduced to 0.8 amperes.

The armature of the machine has a resistance of 0.843 Ω.

2 A 480 volt shunt connected motor draws a total current of 3.6 amperes on no load and 64 amperes on full load. The field circuit resistance is 300 Ω and the armature circuit resistance is 0.25 Ω (excluding brushes). (a) If the armature reaction effect at full load reduces the flux per pole by 1.6% compared to its no load value find the percentage change in speed from no load to full load. (b) What would be the change in speed if the flux had remained constant?

3 A d.c. series motor draws a current of 320 amperes from a 720 volt supply when delivering full load torque at 1000 rev/min. The field and armature resistances are 0.12 Ω and 0.1 Ω, respectively. Find the line current and steady state speed of the machine when: (a) it is delivering half load torque; (b) it delivers full load torque with a divider resistance of 1 Ω in parallel with the field winding; and (c) it delivers full load torque with no diverter resistance but with the voltage reduced to 660 volts.

4 A d.c. series motor operates at 750 rev/min with a line current of 80 amperes from a 230 volt d.c. mains. Its armature resistance is 0.14 Ω and the field resistance is 0.11 Ω. Assuming that the flux corresponding to a current of 20 amperes is 40% of that at a current of 80 amperes, find the motor speed at a line current of 20 amperes when connected to the 230 volt supply.

5 A 4.5 kW, 125 volt, 1150 rev/min, separately excited d.c. generator has an armature circuit resistance of 0.37 Ω. When the machine is driven at rated speed, the open circuit characteristic is given by:

I_f	0.0	0.1	0.2	0.4	0.6	0.8	1.0	1.2	1.4
E_a	4	8	15	29	43	57	72	86	99

I_f	1.6	1.8	2.0	2.2	2.4	2.6	2.8	A
E_a	110	120	126	131	135	138	140	V

The field resistance is adjusted to give a current of 2.0 amperes and the machine driven at 1000 rev/min. (a) What will be the terminal voltage when the machine supplying rated current to a load? (b) If the machine is now driven at 1200 rev/min, what value of field current will be required to deliver rated current at rated voltage?

6 The machine of Exercise 4.5 is now connected as a self excited generator with a field winding resistance of 63 Ω and is driven at rated speed. Sketch the full curve of terminal voltage against load current under these conditions and hence or otherwise find the maximum load current that the machine can supply.

7 A d.c. machine is rated at 10 kW at 120 volts and is to be started by a multistep starter such that the maximum armature current does not exceed 2.4 times the rated current with switching taking place when the armature current has fallen to 1.2 times the rated current. Estimate the number of steps and the value of each step of the starter. The machine has an armature resistance of 0.15 Ω.

5 Three-phase synchronous machines

A three-phase synchronous machine carries a field winding supplied with direct current on its rotor and a three-phase armature winding on its stator. The interaction between the magnetic field produced by the rotor winding and that produced by the stator windings will result in a torque on the rotor which tries to align the rotor field with the stator field. If the fields due to the rotor and stator respectively are rotating in the airgap of the machine at different angular velocities a torque is produced which varies with time and which has a mean value of zero. Thus, in order for a useful torque to be produced, the two fields must be synchronized and hence the speed of rotation of the rotor field (and hence the rotor) must be the same as that of the roating field produced by the stator winding, i.e. at synchronous speed.

Comment

Synchronous speed can be expressed in a variety of ways as follows:

$$n_{sm} = \frac{120f}{p} \qquad \text{synchronous speed in rev/min where } p = \text{number of poles}$$

$$\omega_{sm} = \frac{2\pi}{60} \cdot n_{sm} \qquad \text{synchronous speed in mechanical radians/s}$$

$$\omega = 2\pi f = \frac{p}{2} \cdot \omega_{sm} \qquad \text{synchronous speed in electrical radians/s}$$

In this chapter the synchronous speed will be expressed in electrical radians/s unless otherwise indicated.

With no mechanical load applied to the rotor then, ignoring losses, the rotor and stator fields would be aligned. When a mechanical load is applied to the rotor the resulting torque causes the rotor to decelerate until the resulting misalignment of the stator and rotor fields produces the torque necessary to balance the applied load. This is the motoring condition.

With the machine operating at synchronous speed it is also possible to transfer power from the mechanical system to the electrical system. The synchronous machine is now operating as an **alternator** and, in this mode, provides the bulk of the electricity generated in power systems with machines rated in excess of 1000 MW in use in some power stations.

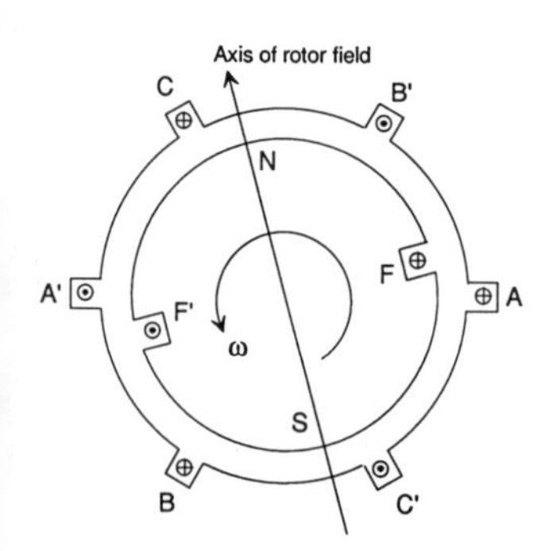

Fig. 5.1 Schematic of the construction of a two-pole, cylindrical rotor synchronous machine.

5.1 Cylindrical rotor machines

Figure 5.1 shows a simplified structure of a two-pole, cylindrical rotor synchronous machine with a d.c. field winding on the rotor and a three-phase armature winding on the stator. Ignoring losses along with any slot effects and assuming a constant reluctance for the flux paths the voltage and torque equations for the machine can be determined as follows.

5.1.1 *Open circuit voltage equation*

With the armature winding open circuit the fundamental component of rotor flux distribution produced by the rotor field winding in the air gap is given by:

$$\boldsymbol{B}_{\mathrm{f}} = \hat{B}_{\mathrm{f}} \,.\, \sin\theta \tag{5.1}$$

Now, using the relationship $e = \boldsymbol{Blv}$, the voltage $\boldsymbol{e}_{\mathrm{a}}$ induced into a single armature conductor at position θ for a rotor of length l, radius r rotating at angular velocity of ω_{sm} is

$$e_{\mathrm{a}} = \boldsymbol{l} \,.\, \omega_{\mathrm{sm}} \,.\, r\hat{B}_{\mathrm{f}} \,.\, \sin\theta \tag{5.2}$$

In terms of the electrical supply frequency since $\omega_{\mathrm{sm}} = 2\omega/p$, where p is the number of poles on the machine the voltage per conductor equation (5.2) becomes:

$$e_{\mathrm{a}} = \frac{2 \,.\, l \,.\, \omega \,.\, r}{p} \hat{B}_{\mathrm{f}} \,.\, \sin\theta \tag{5.3}$$

The r.m.s. voltage (E_{a}) induced in a total of $2N$ conductors in series, representing a coil of N turns, is then

$$E_{\mathrm{a}} = \frac{2 \,.\, \sqrt{2} \,.\, N \,.\, \hat{B}_{\mathrm{f}} \,.\, l \,.\, \omega \,.\, r}{p} \tag{5.4}$$

Taking into account the winding factor (k) for the machine, equation (5.4) becomes

$$E_{\mathrm{a}} = \frac{2 \,.\, \sqrt{2} \,.\, \mathrm{k} \,.\, N \,.\, \hat{B}_{\mathrm{f}} \,.\, l \,.\, \omega \,.\, r}{p} \tag{5.5}$$

Comment

The winding factor (k) is made up of the pitch factor (k_{p}) and distribution factor (k_{d}) which themselves relate to the type of winding used in the machine and thus to the way in which the conductors are distributed around the surface of the machine.

Thus $\quad \mathrm{k} = \mathrm{k}_{\mathrm{p}} \,.\, \mathrm{k}_{\mathrm{d}}$

For a fuller discussion of pitch and distribution factors see O'Kelly (1991).

The fundamental component of flux per pole ($\boldsymbol{\Phi}_{\mathrm{p}}$) can be obtained in terms of the average flux density ($\boldsymbol{B}_{\mathrm{fav}}$) and hence the peak flux density ($\hat{B}_{\mathrm{f}}$) as

$$\boldsymbol{\Phi}_{\mathrm{p}} = \boldsymbol{B}_{\mathrm{fave}} \,.\, \frac{4\pi r l}{p} = \frac{2}{\pi} \hat{B}_{\mathrm{f}} \frac{2\pi r l}{p} = \frac{4rl}{p} \hat{B}_{\mathrm{f}} \tag{5.6}$$

When

$$\hat{B}_{\mathrm{f}} = \frac{p}{4rl} \boldsymbol{\Phi}_{\mathrm{p}} \tag{5.7}$$

Hence, by substituting in equation (5.5)

$$E_a = \left(\frac{1}{\sqrt{2}}\right) . \boldsymbol{\Phi}_p . N . \omega . k = 4.44\ \boldsymbol{\Phi}_p . N . f . k \quad (5.8)$$

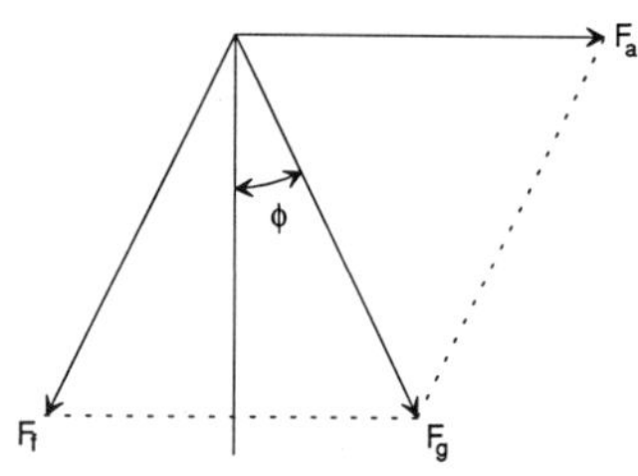

Fig. 5.2 m.m.f. vector diagram for a cylindrical rotor synchronous machine.

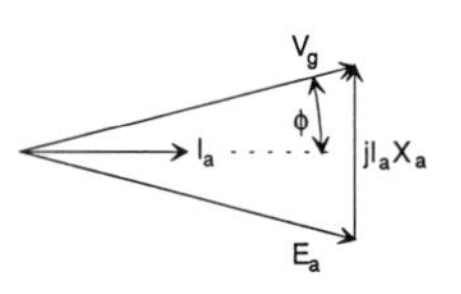

Fig. 5.3 Phasor diagram for the air gap of a cylindrical rotor synchronous machine.

5.1.2. *Equivalent circuit*

Equation (5.8) defines the voltage that will appear at the open circuited armature terminals of a synchronous machine as a result of the rotating field produced by the field winding on the rotor. If a voltage ($\boldsymbol{V}$) is applied to the armature, the resulting current $\boldsymbol{I}_a$ in the armature windings will result in an **armature reaction m.m.f.** in phase with $\boldsymbol{I}_a$.

Ignoring resistance effects the resulting spatial m.m.f. distribution is as illustrated by the m.m.f. vector diagram of Fig. 5.2 in which $\boldsymbol{F}_g$ is the resultant m.m.f. in the airgap, $\boldsymbol{F}_f$ the field m.m.f. and $\boldsymbol{F}_a$ is the armature reaction m.m.f.

The corresponding phasor diagram ignoring resistance is shown in Fig. 5.3 with the armature current $\boldsymbol{I}_a$ as reference. From this diagram it can be seen that effect of armature reaction is to produce a voltage difference between the open circuit voltage ($\boldsymbol{E}_a$) and the air-gap voltage ($\boldsymbol{V}$) which is orthogonal to and proportional to the current ($\boldsymbol{I}_a$) such that for one phase:

$$\boldsymbol{V} = \boldsymbol{E}_a + \boldsymbol{I}_a . jX_a \quad (5.9)$$

where X_a is the armature reactance of the machine. If the stator winding resistance R_a and the stator leakage reactance X_1 are taken into account, the expression for the terminal voltage of one phase becomes:

$$\boldsymbol{V} = \boldsymbol{E}_a + \boldsymbol{I}_a . [R_a + j(X_a + X_1)] \quad (5.10)$$

The armature reactance and stator leakage reactance may be combined to give the synchronous reactance X_s of the machine such that:

$$X_s = X_a + X_1 \quad (5.11)$$

The expression for voltage of equation (5.10) then becomes that of equation (5.12) in which $\boldsymbol{Z}_s$ is the **synchronous impedance** of the machine.

$$\boldsymbol{V} = \boldsymbol{E}_a + \boldsymbol{I}_a . \boldsymbol{Z}_s \quad (5.12)$$

The per-phase equivalent circuit for the synchronous machine is therefore as shown in Figs 5.4(a) and 5.4(b) for motoring and generating conditions respectively.

In practice, and particularly for large machines, the effect of the stator winding resistance R_a can be neglected when:

$$\boldsymbol{V} = \boldsymbol{E}_a + \boldsymbol{I}_a . jX_s \quad (5.13)$$

Comment

The equivalent circuit assumes a star connected machine. If the machine is connected in delta then the equivalent star connection must be determined.

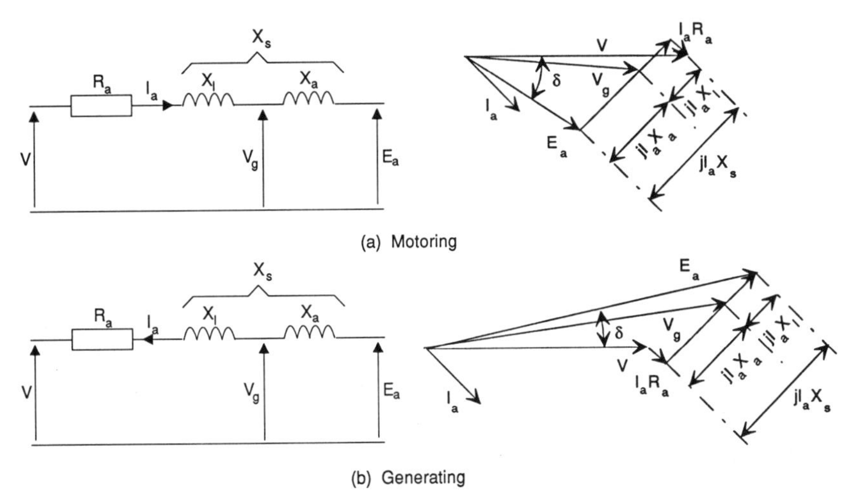

Fig. 5.4 Per-phase equivalent circuit of a synchronous machine: (a) motoring; and (b) generating.

The angle δ between the applied armature voltage ($\boldsymbol{V}$) and the open-circuit voltage ($\boldsymbol{E}_a$) is referred to as the **rotor angle** or **machine angle** and is taken to be positive when $\boldsymbol{V}$ leads $\boldsymbol{E}_a$, as is the case in Fig. 5.5 where the machine is motoring.

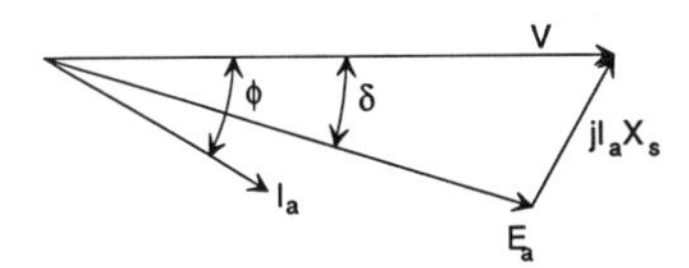

Fig. 5.5 Phasor diagram for one phase of an ideal synchronous machine when motoring.

Comment

- When motoring the rotor field will lag the stator field by an angle δ.
- When generating the rotor field will lead the stator field by an angle δ.

5.1.3 *Torque equation*

For an ideal synchronous machine with negligible stator resistance the electrical power at the machine terminals must be the same as the mechanical shaft power. Hence, referring to the equivalent circuit:

$$P_{\text{elec}} = 3 \,.\, V \,.\, I_a \,.\, \cos\phi = T \,.\, \omega_{\text{sm}} = P_{\text{mech}} \tag{5.14}$$

when

$$T = \frac{3 \,.\, V \,.\, I_a \,.\, \cos\phi}{\omega_{\text{sm}}} = \left(\frac{p}{2 \,.\, \omega}\right) . \, 3 \,.\, V \,.\, I_a \,.\, \cos\phi \tag{5.15}$$

Now, from Fig. 5.5:

$$I_a \,.\, \cos\phi = \frac{E_a \,.\, \sin\delta}{X_s} \tag{5.16}$$

Hence

$$P_{\text{elec}} = \frac{3 \,.\, V \,.\, E_a \,.\, \sin\delta}{X_s} \tag{5.17}$$

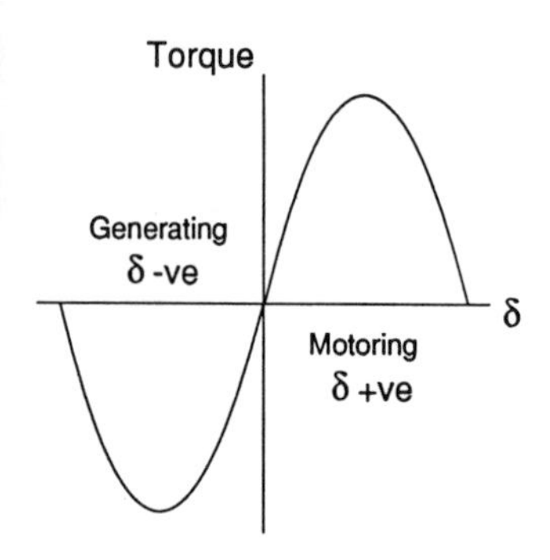

Fig. 5.6 Torque characteristic of a cylindrical rotor synchronous machine.

or in terms of line voltages:

$$P_{\text{elec}} = \frac{V_{\text{line}} \, . \, E_{\text{aline}} \, . \sin \delta}{X_s} \tag{5.18}$$

and

$$T = \left(\frac{p}{2 \, . \, \omega}\right) . \frac{V_{\text{line}} \, . \, E_{\text{aline}} \, . \sin \delta}{X_s} \tag{5.19}$$

The torque characteristic for the cylindrical rotor synchronous machine is therefore as shown in Fig. 5.6.

Comment

The maximum torque that a synchronous machine can deliver when motoring or absorb when generating occurs when $\delta = 90^\circ$ and is referred to as the pull-out torque. Any attempt to increase the torque beyond this value will result in a loss of synchronism as for $\delta > 90^\circ$ the torque available decreases with increasing values of δ.

The pull-out torque is usually several times greater than the rated torque of the machine.

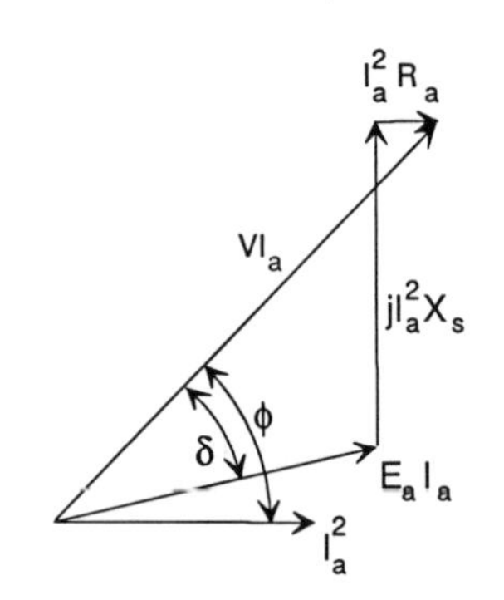

Fig. 5.7 Power diagram for a three-phase cylindrical rotor synchronous machine when motoring.

When motoring, the useful mechanical torque is obtained by subtracting the rotational losses from the developed torque:

$$P_{\text{out}} = \frac{V_{\text{line}} \, . \, E_{\text{aline}} \, . \sin \delta}{X_s} - P_{\text{loss}} = P_{\text{elec}} - P_{\text{loss}} \tag{5.20}$$

If the resistance of the machine windings is taken into account then, referring to Fig. 5.7 for motoring operation:

$$P_{\text{elec}} = 3 \, . \, V \, . \, I_a \, . \cos \phi - 3 \, . \, I_a^2 \, . \, R_a = 3 \, . \, E_a \, . \, I_a \, . \cos(\phi - \delta) \tag{5.21}$$

and

$$T = \left(\frac{p}{2 \, . \, \omega}\right) . \, (3 \, . \, V \, . \, I_a \, . \cos \phi - 3 \, . \, I_a^2 \, . \, R_a) = \left(\frac{p}{2 \, . \, \omega}\right) . \, 3 \, . \, E_a \, . \, I_a \, . \cos(\phi - \delta) \tag{5.22}$$

Comment

For the synchronous machine when generating (energy flow from the mechanical system to the electrical system) the corresponding relationships are as follows.

Machine equations per phase:

Ideal machine $\quad \boldsymbol{E}_a = \boldsymbol{V} + \boldsymbol{I}_a \, . \, jX_s$

Including resistance $\quad \boldsymbol{E}_a = \boldsymbol{V} + \boldsymbol{I}_a \, . \, (R_a + jX_s)$

Rotational losses are provided by the mechanical source when, ignoring resistance:

$$P_{\text{in}} = \frac{V_{\text{line}} \, . \, E_{\text{aline}} \, . \sin \delta}{X_s} + P_{\text{loss}} = P_{\text{elec}} + P_{\text{loss}}$$

or, taking into account rotor resistance and referring to Fig. 5.8:

$$P_{\text{elec}}=3\,.\,V\,.\,I_a\,.\cos\phi+3\,.\,I_a^2\,.\,R_a=3\,.\,E_a\,.\,I_a\,.\cos(\phi+\delta)$$

and

$$T=\left(\frac{p}{2\,.\,\omega}\right).\,(3\,.\,V\,.\,I_a\,.\cos\phi+3\,.\,I_a^2\,.\,R_a)=\left(\frac{p}{2\,.\,\omega}\right).\,3\,.\,E_a\,.\,I_a\,.\cos(\phi+\delta)$$

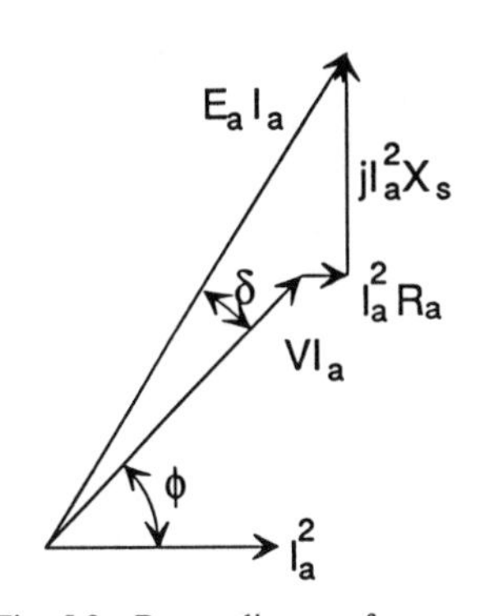

Fig. 5.8 Power diagram for a three-phase cylindrical rotor synchronous machine when generating.

Example 5.1 Synchronous machine performance

A 660 V, 50 Hz, three-phase, four-pole, star connected cylindrical rotor synchronous generator has an armature resistance of 0.42 Ω/phase and a synchronous reactance of 7.2 Ω/phase. Find the mechanical input power and torque required when it is operating as a generator supplying a line current of 24 amperes at a lagging power factor of 0.92. The mechanical losses of the machine (windage and friction) are 1.2 kW.

What would be the effect on the calculated torque of ignoring the winding resistance?

Taking the phase voltage V as reference.

$$\boldsymbol{I}_a=24\angle-23.07^\circ\ \text{A}$$

and

$$\boldsymbol{V}=381\angle 0^\circ\ \text{V}$$

Now

$$\boldsymbol{Z}_s=0.42+7.2\text{j}=7.21\angle 86.66^\circ\ \Omega$$

Hence

$$\boldsymbol{E}_a=\boldsymbol{V}+\boldsymbol{I}_a\,.\,\boldsymbol{Z}_s=457.9+154.8\text{j}=483.4\angle 18.68^\circ=483.4\angle\delta\ \text{V}$$

Thus, from above:

$$P_{\text{elec}}=3\,.\,E_a\,.\,I_a\,.\cos[(23.07+18.68)^\circ]=25.97\ \text{kW}$$

Total input power (P_{mech}) is then

$$P_{\text{mech}}=P_{\text{elec}}+\text{mechanical losses}=27.17\ \text{kW}$$

Hence

$$T=\left(\frac{p}{2\,.\,\omega}\right)\times 27.17\times 10^3=173\ \text{N m}$$

Ignoring rotor resistance

$$\boldsymbol{E}_a=\boldsymbol{V}+\boldsymbol{I}_a\,.\,X_s=448.7+159\text{j}=476\angle 19.51^\circ$$

$$P_{\text{elec}}=\frac{3\,.\,V\,.\,E_a\,.\sin\delta}{X_s}=\frac{3\times 381\times 483.4\times\sin(19.51^\circ)}{7.2}=25.63\ \text{kW}$$

when

$$T=170.8\ \text{N m}$$

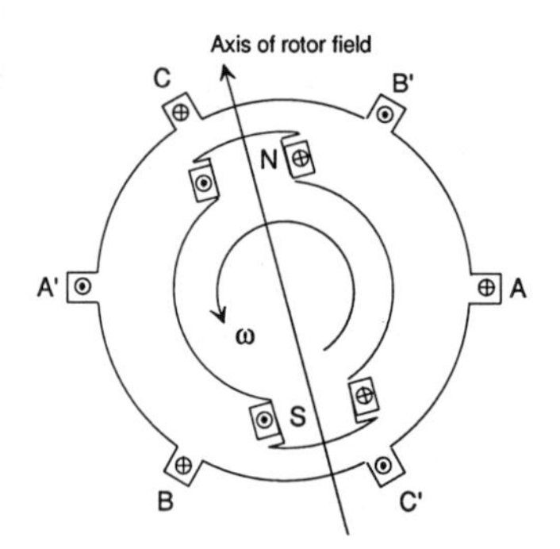

Fig. 5.9 Two-pole salient pole machine.

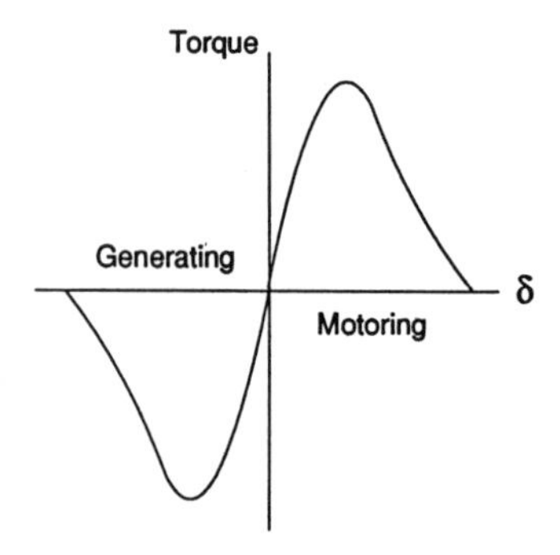

Fig. 5.10 Torque characteristic of a salient pole synchronous machine.

5.2 Salient-pole synchronous machines

The construction of a simple two-pole salient-pole synchronous machine is shown in Fig. 5.9. The effect of the asymmetry introduced by the field poles is to introduce a saliency or reluctance torque of the form:

$$T_r = k_s \,.\, V^2 \,.\, \sin 2\delta \tag{5.23}$$

where k_s is a constant determined by the construction of the machine. This has the effect of modifying the torque characteristic to that of Fig. 5.10.

Comment

The torque characteristic of Fig. 5.10 contains a fundamental component together with a second harmonic as a result of the presence of saliency. See Chapter 3 for a fuller discussion of saliency torque.

5.3 Synchronous machine excitation

The operation of a synchronous machine in both motoring and generating modes is influenced by the strength of the field and hence by the magnitude of the induced or open circuit voltage E_a. Three possible conditions exist.

- Underexcitation: the magnitude of the open circuit voltage (E_a) is less than that of the terminal voltage (V) ($E_a < V$).
- Level excitation: E_a and V are equal in magnitude ($E_a = V$).
- Overexcitation: the magnitude of E_a is greater than that of V(E_a > V).

Comment

The armature voltage equations for the ideal cylindrical rotor synchronous machine are:

$$V = E_a + jI_a \,.\, X_s \qquad \text{Motoring}$$

$$E_a = V + jI_a \,.\, X_s \qquad \text{Generating}$$

The operation of an ideal cylindrical rotor synchronous machine in the motoring mode for each of these conditions is illustrated by the phasor diagrams of Figs 5.11(a), 5.12(a) and 5.13(a), respectively while the corresponding phasor diagrams for the generating mode are given by Figs 5.11(b), 5.12(b) and 5.13(b).

Referring to Fig. 5.11 it is seen that an underexcited synchronous machine will operate with a lagging power factor (I_a lags V) when motoring and with a leading power factor (I_a leads V) when generating.

Referring now to Fig. 5.13 for the overexcited synchronous machine it is seen that the

Fig. 5.11 Synchronous machine – underexcited operation: (a) motoring; and (b) generating.

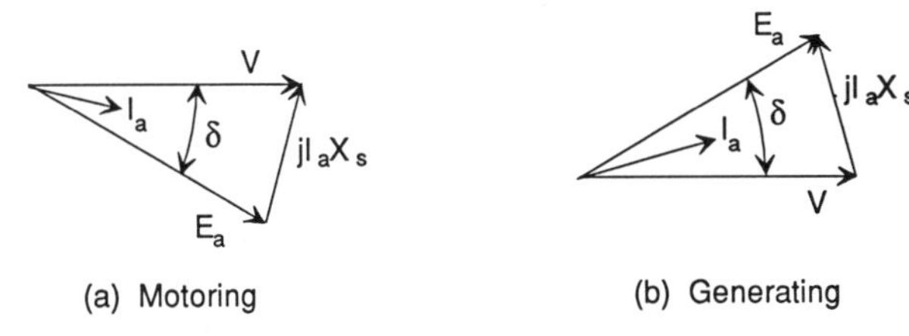

Fig. 5.12 Synchronous machine – level excitation: (a) motoring; and (b) generating.

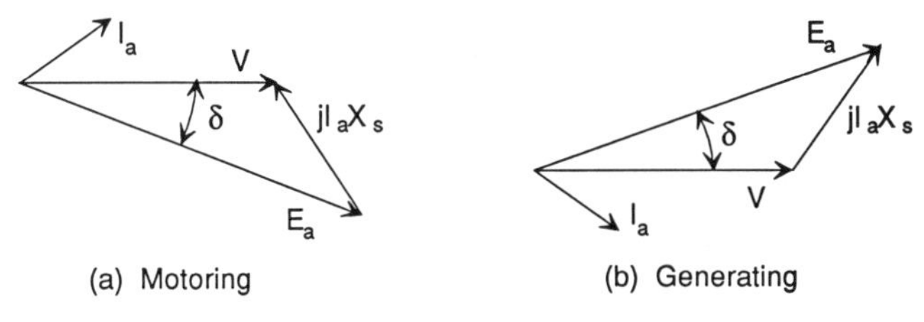

Fig. 5.13 Synchronous machine – overexcited operation: (a) motoring; and (b) generating.

situation is now reversed and the machine operates with a leading power factor (I_a leads V) when motoring and with a lagging power factor (I_a lags V) when generating.

Comment:

With reference to Chapter 2 and the discussion of real and reactive power it can be seen that an underexcited synchronous machine in motoring mode will appear as an inductive load and absorb lagging VAR while in the generating mode it delivers leading VAR.

Similarly, an overexcited synchronous machine in motoring mode will appear as a capacitor and absorb leading VAR while in the generating mode it delivers lagging VAR.

This ability of the synchronous motor to appear on the system as a capacitive load means that they can be used as an alternative to capacitor banks as a source of lagging VAR for power factor correction and in association with controlled converters for high-voltage d.c. (HVDC) transmission systems and similar applications. When operated in this way the machine would be referred to as a **synchronous compensator** or a **synchronous condensor**.

Comment

- For power factor correction see Chapter 2, section 2.
- For an introduction to the operation of controlled converters see Chapter 3.

Example 5.2 Synchronous machine excitation
An 11 kV, three-phase, star connected synchronous machine with a synchronous reactance of 7.8 Ω/phase and is being used as an alternator delivering 200 A to an 11 kV power system at a lagging power factor of 0.9. If the mechanical power supplied to the machine is maintained constant, what will be the new current and power factor if the excitation of the machine is (a) increased by 20% and (b) decreased by 10%?

Taking the terminal voltage $\boldsymbol{V}$ as reference:
Initially

$$\boldsymbol{I}_a = 200\underline{|-25.84^\circ} = 180 - 87.2\text{j A}$$

and

$$\boldsymbol{E}_a = \boldsymbol{V} + \boldsymbol{I}_a \,.\, \text{j}X_s = 7031 + 1404\text{j} = 7169.8\underline{|11.29^\circ}\text{ V}$$

Input power

$$P_{elec} = \frac{3\,.\,E_a\,.\,\text{V}\,.\,\sin\delta}{X_s} = 3.429\text{ MW}$$

(a) After increase in excitation

$$E_a = 1.2 \times 7169.8 = 8603.8\text{ V}$$

as power is constant

$$\delta = \sin^{-1}\left(\frac{P_{elec}\,.\,X_s}{3\,.\,E_a\,.\,V}\right) = 9.39^\circ$$

Therefore, $\boldsymbol{E}_a$ leads $\boldsymbol{V}$ by δ when with $\boldsymbol{V}$ as reference:

$$\boldsymbol{I}_a = \frac{\boldsymbol{E}_a - \boldsymbol{V}}{\text{j}X_s} = \frac{8603.8\underline{|9.39^\circ} - 6350.6\underline{|0^\circ}}{7.8\underline{|90^\circ}} = 327.9\underline{|-56.08^\circ}\text{ A}$$

For which the power factor is 0.558 lagging.

(b) After reduction in excitation

$$E_a = 0.9 \times 7169.8 = 6452.8\text{ V}$$

as power is constant

$$\delta = \sin^{-1}\left(\frac{P_{elec}\,.\,X_s}{3\,.\,E_a\,.\,V}\right) = 12.56^\circ$$

Therefore, $\boldsymbol{E}_a$ leads $\boldsymbol{V}$ by δ when with $\boldsymbol{V}$ as reference:

$$\boldsymbol{I}_a = \frac{\boldsymbol{E}_a - \boldsymbol{V}}{jX_s} = 180\underline{|2.15^\circ}$$

For which the power factor is 0.9993 leading

5.4 Synchronous machine stability

An increase in the load on a synchronous machine will result in a deceleration of the rotor to establish the new, and higher, value of the power angle (δ). In practice, the rotor will **overshoot** the required power angle and will then hunt around this position in a manner determined by the amount of damping present in the system. If the change in load is too great, it is possible that synchronism could be lost. The **equal area criterion** is used to define the limits for step changes in load for an undamped machine.

5.4.1 *Equal area criterion*

Consider a situation where the load torque on the synchronous machine is increased from T_1 to T_2. Referring to Fig. 5.14, this would correspond to a change in the power angle from δ_1 to δ_2. However, in the absence of any damping, when the power angle (δ) reaches δ_2, the rotor has a negative velocity relative to synchronous speed and hence δ will continue to increase. In moving from δ_1 to δ_2 the rotor of the machine has given up an amount of energy represented by the area A_1 in Fig. 5.14. For values of δ greater than δ_2, the machine is developing an excess of torque with respect to the load torque, bringing the rotor to rest, relative to synchronous speed, at δ_3. When δ reaches δ_3, the energy lost as δ increased from δ_1 to δ_2 has been regained. This energy is represented by the area A_2 in Fig. 5.14 when, referring to Fig. 5.14:

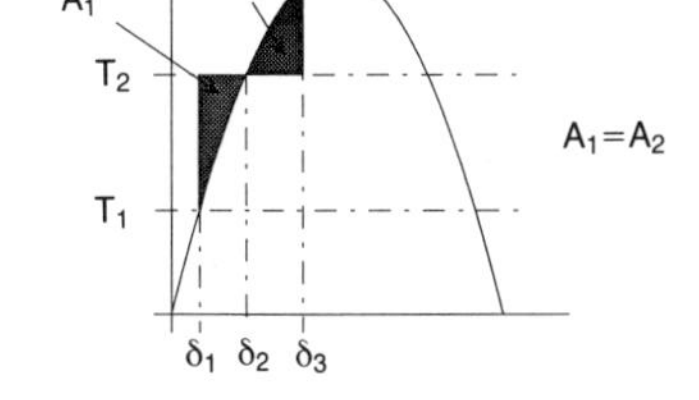

Fig. 5.14 Equal area criterion.

$$A_1 = A_2 \tag{5.24}$$

when:

$$\int_{\delta_1}^{\delta_2} (\sin\delta_2 - \sin\delta)\,.\,d\delta = \int_{\delta_2}^{\delta_3} (\sin\delta - \sin\delta_2)\,.\,d\delta \tag{5.25}$$

the limiting condition for which stability is just maintained is shown in Fig. 5.15.

In the absence of damping, δ would now oscillate continuously between δ_1 and δ_3.

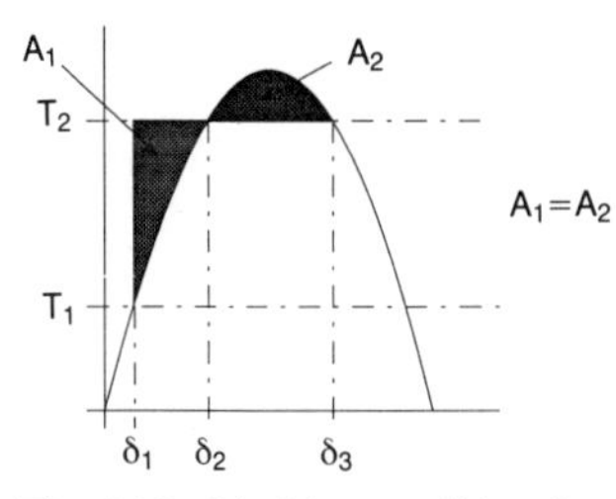

Fig. 5.15 Limiting condition for equal area criterion.

Comment

Solving equation (5.25) gives:

$$\delta_1\,.\,\sin\delta_2 + \cos(\delta_1) = \delta_3\,.\,\sin\delta_2 + \cos(\delta_3)$$

Example 5.3 Equal area criterion
A synchronous motor is operating at 10% of maximum torque when a step change of load equivalent to 15% of maximum torque is applied. Determine if the machine will

remain in synchronism and what will be the maximum value of rotor angle reached during hunting.

What would be the maximum step change of torque that could be applied if synchronism was to be maintained? Damping effects may be neglected.

(a) For a torque of 10% of T_{max}:

$$\delta_1 = \sin^{-1}(0.1) = 5.74° = 0.1002 \text{ radians}$$

At the new load torque of 25% of T_{max}:

$$\delta_2 = \sin^{-1}(0.25) = 14.48° = 0.2527 \text{ radians}$$

Therefore, from the solution of equation (5.25) above:

$$0.25 \times (0.2527 - 0.1002) - 0.995 = 0.25 \times (0.2527 - \delta_3) - \cos \delta_3$$

When

$$1.02 = 0.25\,\delta_3 + \cos \delta_3$$

Solving:

$$\delta_3 = 0.408 \text{ radians} = 23.38°$$

(b) At the limit of stability:

$$\delta_3 = \pi - \delta_2$$

When:

$$0.995 = \sin \delta_2(3.041 - \delta_2) + \cos \delta_2$$

Solving:

$$\delta_2 = 0.8562 \text{ radians} = 49.06°$$

This corresponds to a load of $(100 \times \sin \delta_2)\% \cong 75.5\%$ of full load or a step increase in load of 65.5%.

(a) *Damper windings*

The equal area criterion defines a worst case condition for an undamped machine. In a practical machine, some mechanical damping will always be present as a result of windage and friction and, for small machines, this may be sufficient in its own right. Where additional damping is required, short circuited **damper windings** may be included in the surface of the rotor. During the hunting period, the relative motion between these conductors and the field induces a current in the damper windings which results in a torque opposing and damping the relative motion of the rotor.

5.5 Excitation systems

5.5.1 *d.c. exciter*

Prior to the availability of high-power diodes, the field winding of a large synchronous machine would have been supplied via slip-rings by a **d.c. exciter**, probably running or

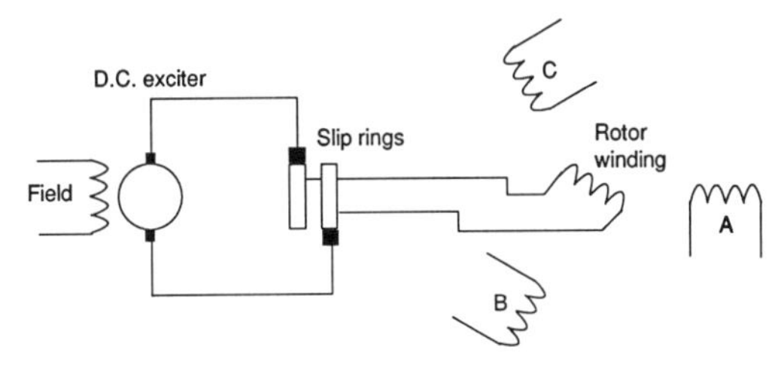

Fig. 5.16 d.c. exciter.

the same shaft as the synchronous machine, as in Fig. 5.16. Though effective, this system requires a significant maintenance input, particularly for the brush gear on the exciter and slip-rings.

5.5.2 *Brushless exciter*

In a **brushless exciter** the need for brushgear is eliminated by introducing a diode rectifier mounted on the machine shaft and connected to an a.c. exciter, also on the same shaft. Typical arrangements for motors and generators respectively are shown in Figs 5.17(a) and (b).

In the case of the motor on start-up the voltage induced in the field circuit will be large and this is accommodated by the zener diode which conducts under these conditions, inserting the damping resistor into the field circuit. As the speed builds up, this voltage falls to a point at which the control circuit cuts in the field and the machine synchronizes.

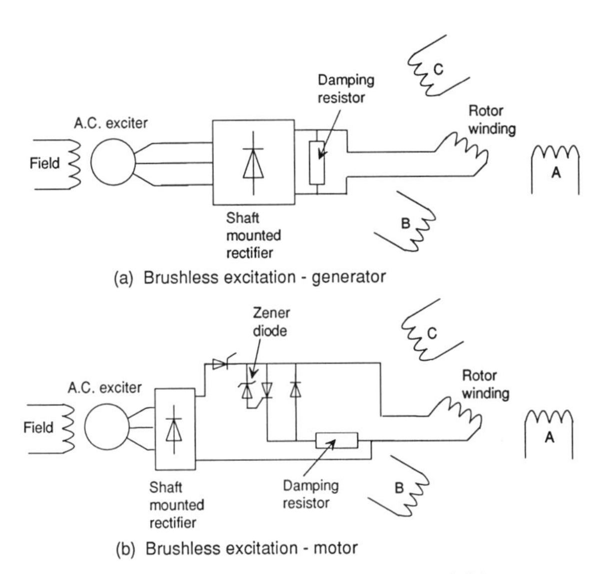

Fig. 5.17 Brushless exciters: (a) generator; and (b) motor.

5.5.3 *Permanent magnet machines*

The availability of high-strength permanent magnets has made available a range of machines in which the field is supplied by a permanent magnet on the rotor. These machines have found a wide range of applications, particularly as servomotors in control systems. The field of a permanent magnet machine is fixed and ratings are limited to a few kVA by magnet performance and cost.

Comment

Permanent magnet machines are also referred to as brushless d.c. machines (Chapter 4). The actual operating mode depends upon the type of control used.

5.6 Synchronous machine starting

As has been shown, a synchronous machine only develops torque at synchronous speed and does not have any intrinsic starting torque capability and a means of accelerating from rest to synchronous speed must be provided. Once at or very near synchronous speed the field can then be applied and the rotor will pull in and the load can be applied.

Where a synchronous motor is fitted with a d.c. exciter on the same shaft this exciter can be used as a motor to drive the combination up to synchronous speed. Where the motor is fitted with damper windings to control hunting these can be used to run the motor up as an induction motor on no-load to near synchronous speed when application of the field will pull it into synchronism.

Comment

See Chapter 6 for the discussion of induction motors.

In situations where the synchronous motor is supplied from a variable frequency inverter then acceleration and run-up can be controlled by varying the frequency of the supply.

Comment

See Chapter 2 for the discussion of inverters.

In the case of a generator the prime mover is used to run the machine up to, or near synchronous speed on open circuit. The field is then applied to the machine and i adjusted so that the open circuit voltage matches that of the supply to which the machine is to be connected. Providing the phase sequence of the generator matche that of the supply the switch connecting the machine to the supply can then be closed

5.7 Synchronous machine operation

Assuming a constant-frequency, constant-voltage source, the operation of the synchronous machine can be summarized by Fig. 5.18(a) using the convention that a lagging power factor load absorbs reactive power.

Referring to Fig. 5.18(b), the effect of varying the power factor on machine performance can be seen. Assuming constant power at the machine terminals then $I_a \cos \phi$ is constant and the locus of I_a follows the path AA′. Under these conditions, the locus of E_a is along the line BB′ parallel to the supply voltage axis.

If instead the excitation is held constant then the magnitude of E_a will remain constant and the locus of E_a will be an arc centred at the origin O as in Fig. 5.18(c).

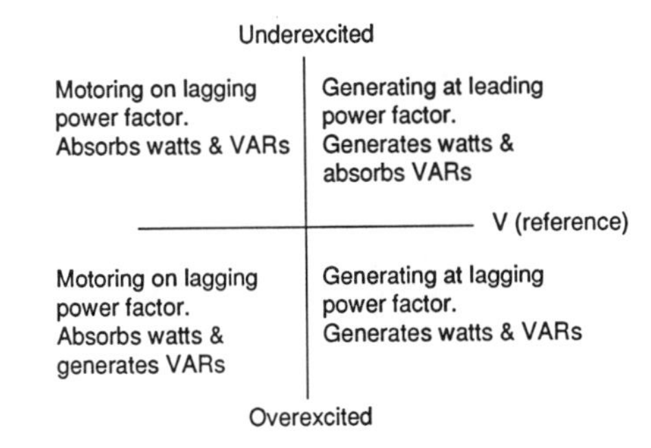

(a) Synchronous machine operation

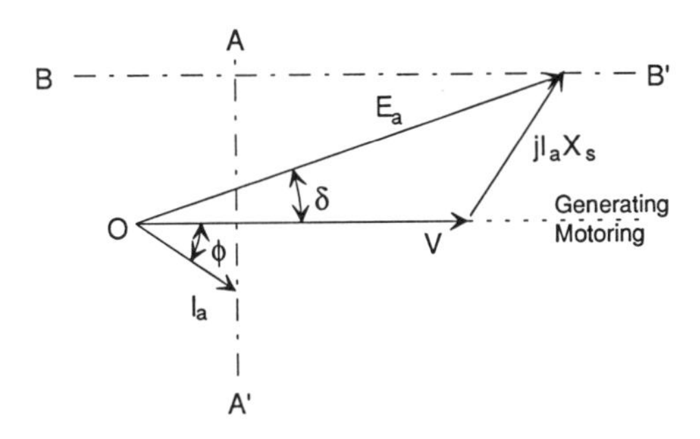

(b) Constant power operation

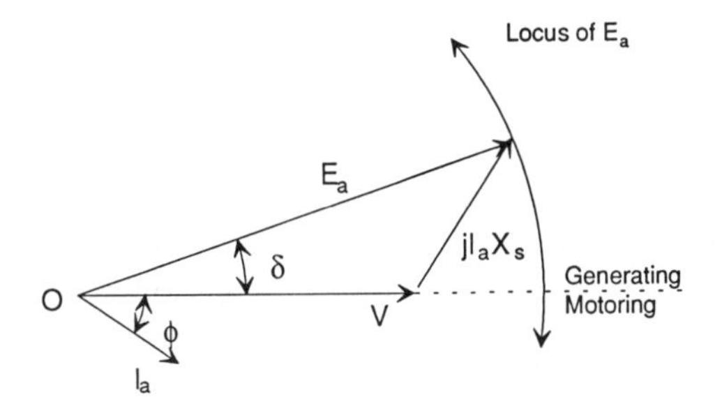

(c) Operation with constant excitation

Fig. 5.18 (a) Synchronous machine operation, (b) constant power operation and (c) operation with constant excitation.

Operating limits can be included on Fig. 5.18 as follows.

- The maximum power of the machine. This is a straight line parallel to the VAR axis representing the maximum power that the prime mover driving the generator is capable of supplying.
- The maximum VA of the machine. This is a circle centred at point A, the radius of which is defined by the product of the busbar voltage and the rated armature current ($3VI_{\text{rated}} = \sqrt{3}V_{\text{line}}I_{\text{rated}}$).
- The maximum rotor loss conditions for the machine. This a circle centred at O, the radius of which is determined by the maximum field current and hence the maximum value of E_a.
- The practical stability limit for the machine. The theoretical limit of stability of the machine corresponds to a power angle (δ) of 90°. The practical stability limit defines the effective limit on power angle under normal operation and is chosen in line with the intended application of the machine.

Comment

A similar diagram could be drawn for the motoring case using:

$$\boldsymbol{V} = \boldsymbol{E}_a + j\boldsymbol{I}_a \cdot X_s$$

5.8 Speed control

The speed control of a three-phase synchronous motor depends upon the provision at the machine terminals of a three-phase voltage of variable magnitude and frequency such that:

$$\frac{V}{f} = \text{constant} \tag{5.26}$$

As both the open circuit voltage (E_a) and machine reactances are proportional to speed and hence to frequency the adoption of the relationship of equation (5.26) means that the operating flux density, and hence the field characteristics, will be substantially constant over the operating speed range of the machine.

5.8.1 *Inverter drive*

The generation by the rotating field of a three-phase voltage in the stator winding of a three-phase synchronous machine is such that it supports the natural commutation of the inverter switching elements and hence a fully controlled, three-phase thyristor bridge operating in inverter mode can be used as the supply to the machine. A typical arrangement is shown in Fig. 5.19 in which the supply converter is used to provide a variable d.c. voltage to the machine converter; rotor position sensing is used to control the firing of the thyristors.

As a bridge converter absorbs reactive power, the current into the machine when motoring must always lead the terminal voltage and the machine operates at leading power factor.

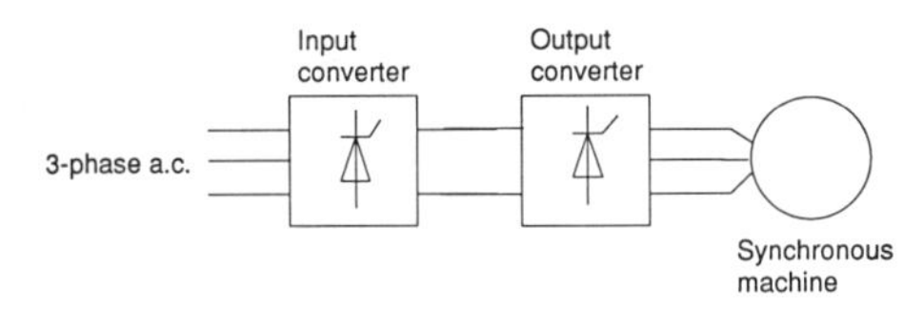

Fig. 5.19 Synchronous machine with inverter drive.

The speed of the machine is effectively determined by the level of the d.c. link voltage when, in the absence of any speed feedback, there will be some speed regulation with increasing torque. The use of both analogue and/or digital controllers does, however, enable a wide range of different control strategies to be implemented.

As both converters are naturally converted, power may flow in either direction enabling regenerative braking to be used to rapidly decelerate the machine with power being returned to the supply. Once zero speed has been reached, the direction of rotation can be reversed by altering the switching sequence of the machine converter to give full four-quadrant operation.

On starting, natural commutation of the machine converter is not possible as no open circuit voltage is being developed. By switching the supply inverter between the rectifying and inverting modes a series of current pulses can be supplied to the machine, producing a starting torque. As the speed of the machine increases the transition to natural commutation takes place at around 20% of nominal synchronous speed.

Example 5.4 Synchronous machine control

A six-pole, three-phase synchronous motor rated at 1.2 MVA at 11 kV and 50 Hz. The machine has negligible stator resistance per phase and a synchronous reactance of 10.2 Ω/phase. It is energized by a three-phase fully controlled thyristor bridge converter operating in inverting mode from a d.c. link supplied by a second, similar converter. When operating with the field current set to give an open circuit voltage equal to 95% of the rated voltage of the machine the inverter operates with an extinction angle (ε) of 12° and has a fundamental component of line current whose r.m.s. value is equal to the rated current of the machine.

Find the d.c. link voltage and the firing advance angle (β) under these conditions.

$$E_a = \frac{0.95 \times 11\,000}{\sqrt{3}} = 6033.5 \text{ V}$$

The r.m.s. value of fundamental component of line current is:

$$I_{a1} = \frac{1.2 \times 10^6}{\sqrt{3} \times 11\,000} = 62.98 \text{ A}$$

Referring to Fig. 5.20

$$\hat{I}_{a1} = \sqrt{2} \,.\, I_{a1} = \frac{1}{\pi}\left(\int_{-\pi}^{\frac{-2\pi}{3}} -I_{dc} \,.\, \cos\theta \,.\, d\theta + \int_{\frac{-\pi}{3}}^{\frac{\pi}{3}} I_{dc} \,.\, \cos\theta \,.\, d\theta + \int_{\frac{-2\pi}{3}}^{\pi} -I_{dc} \,.\, \cos\theta \,.\, d\theta\right) = \frac{I_{dc} \,.\, 2\sqrt{3}}{\pi}$$

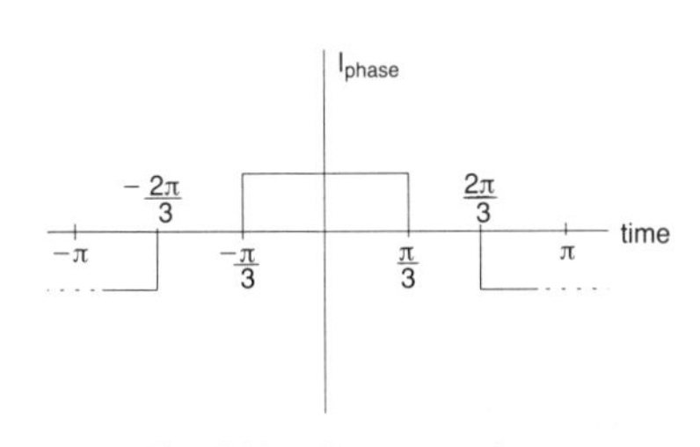

Fig. 5.20 Converter phase current waveform (Example 5.4).

Therefore

$$I_{dc} = \frac{I_{al} \cdot \pi}{\sqrt{6}} = 80.77 \text{ A}$$

Now,

$$\hat{V} = \sqrt{2} \cdot \sqrt{3} \cdot E_a = 14\,779 \text{ V}$$

From Chapter 2 (equations (2.10) and (2.11)):

$$\frac{1}{2} \cdot \hat{V} \cdot (\cos\beta + \cos\varepsilon) = \hat{V} \cdot \cos\beta + I_{dc} \cdot R_i$$

when

$$\cos\beta = 0.8667$$

and

$$\beta = 29.93°$$

The d.c. link voltage is then

$$V_{dc} = \frac{3}{2\pi} \cdot 14779 \cdot [\cos(29.93°) + \cos(12°)] = 13\,018 \text{ V}$$

5.8.2 *Cycloconverter drive*

The primary application of cycloconverters for the control of synchronous machines is in association with large, high-power drives operating at low speeds. Such drives offer high levels of efficiency with good control but tend to have a restricted speed range – from near zero speed to around one-third or one-half of the nominal synchronous speed set by the supply frequency.

5.9 Applications

The major application of large synchronous machines is as an alternator for power generation. A typical alternator would be of two-pole or four-pole construction and driven by a steam or gas turbine with single machines capable of providing in excess of 1000 MW. The rotor radius of such machines is constrained by the need to maintain the slots and windings against the centrifugal forces developed and they therefore tend to be of a long, thin construction with a cylindrical rotor.

On the other hand, for applications such as hydro-power generation the speeds of rotation are much less – of the order of 200 or 300 rev/min – requiring large numbers of poles. Such machines tend to be short with a large radius and of salient pole construction.

Synchronous motors are primarily used to provide constant speed drives, particularly in the higher power (MW) ranges, where they offer a greater efficiency than the equivalently rated d.c. machine or induction machine. Typical applications include large fans and blowers, compressors, mills, conveyors and centrifuges.

Exercises

1 A 1 MW, 50 Hz, three-phase, 11 kV (line) star connected alternator has a resistance per phase of 1 Ω, a core loss of 25 kW, a field copper loss of 6.8 kW and a combined windage and friction loss of 9.2 kW. Find the full load efficiency at a power factor of 0.9 lagging.

2 A 50 Hz, three-phase, four-pole, 2200 volt (line) has a synchronous reactance of 1.64j Ω per phase and is rated at 125 kVA. What will be the armature current and machine voltage when the machine is delivering 100 kVA to a three-phase, 2200 volt (line) busbar at: (a) unity power factor; (b) at leading power factor of 0.8; and (c) a lagging power factor of 0.85?

3 A cylindrical rotor, three-phase synchronous machine is operating as a motor drawing a current of 65.4 amperes per phase from a 415 volt (line) busbar at a leading power factor of 0.88. If the machine has a synchronous reactance of 0.94j Ω, find the machine angle δ and the machine voltage E_a.

4 A three-phase, 50 Hz, 120 MVA, 13.8 kV (line) star-connected alternator has a synchronous reactance of 1.3j Ω per phase and is connected to a constant voltage, constant frequency busbar supplying half load current at a lagging power factor of 0.8. If the excitation is now increased by 20% and the mechanical power input by 50% determine the new operating power factor and machine load angle. Saturation and machine resistance are neglected.

5 A three-phase, two-pole synchronous machine is rated at 0.8 MVA at 6600 volt (line) and 50 Hz and has a synchronous reactance of 48j Ω. The machine is connected to a 6600 volt busbar and is used to supply a proportion of a factory load; the remainder of the load is supplied by a three-phase, 6600 volt (line) transmission line.

(a) If the machine is supplying 30% of the total load of 1.2 MW, the power factor of which is 0.85 lagging, find the current in the transmission line if the synchronous machine is overexcited by 25%.

(b) Under what conditions of machine excitation will the transmission line supply 0.75 MW at a power factor of 0.9 lagging at the busbar to this same load?

6 A synchronous motor is operating at 40% of full load when the load is suddenly increased to 80% of full load, determine if the machine remains in synchronism as a result of this change in load.

6 Induction machines

Induction machines are by far the most numerous of electric machines, accounting for over 80% of the motors in use. In size they range from the 50 watt, **single-phase shaded-shaded pole** motors used for fans through the **single-phase squirrel-cage** motors in domestic appliances such as washing machines and refrigerators to **three-phase** 25 MW machines used in steel works and other industrial applications. Though rarely deployed in a generating mode, **induction generators** have been used in wind generators because of their inherent speed regulating performance.

6.1 Torque production

The torque production in an induction machine relies on the provision of a rotating magnetic field in the airgap of the machine, as in Fig. 6.1(a), by a current (single-phase machine) or currents (three-phase machine) in the stator winding. Assuming the rotor to be stationary initially, this rotating magnetic field induces an e.m.f. in the rotor conductors. When connected to an external circuit this e.m.f. will result in a current in the rotor conductor in the direction shown in Fig. 6.1(b).

This current would then interact with the rotating magnetic field resulting in a force on the conductor, and hence a torque, in the same direction as the direction of rotation of the stator magnetic field, as in Fig. 6.1(c), which will cause the machine to accelerate. In order to maintain the torque against the mechanical losses of the machine (windage and friction) and against any applied load, a relative motion must be maintained between the stator field and the rotor conductors in order that a voltage, and hence a current, can be induced into the rotor conductors.

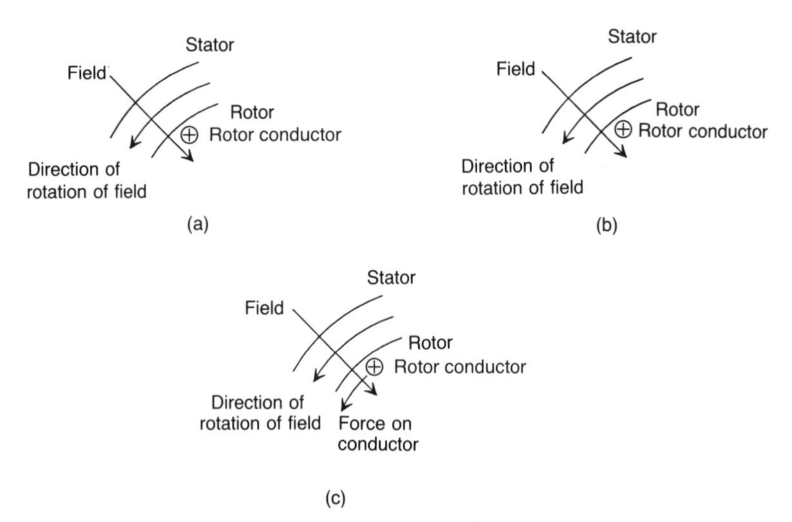

Fig. 6.1 Torque production in an induction machine: (a) airgap of induction machine showing stator field and direction of rotation; (b) direction of current in rotor conductor as a result of stator field; and (c) direction of resulting force on rotor conductor.

This difference between the rotational speed of the stator field in the airgap and the rotational speed of the rotor is expressed by the slip (s) such that:

$$\text{slip} = s = \frac{\text{Rotational speed of stator field in airgap} - \text{Rotational speed of rotor}}{\text{Rotational speed of stator field in airgap}} \tag{6.1}$$

or

$$s = \frac{\text{Synchronous speed} - \text{Actual speed}}{\text{Synchronous speed}} \tag{6.2}$$

and

$$\text{Synchronous speed} = \frac{120\,.\,f}{p}\ \text{rev/min} = \frac{2}{p}\,.\,\omega\ \text{radians/s} = \omega_{sm}\ \text{radians/s} \tag{6.3}$$

where:

f is the supply frequency;
p is the number of poles on the machine;
ω_{sm} is the synchronous speed in mechanical radians/s.

Comment

The equation for slip may be expressed in a variety of ways as shown below in which:

n_m is the speed of the machine in rev/min;
n_{sm} is the synchronous speed of the machine in rev/min;
ω_m is the speed of the machine in mechanical radians/s;
ω_{sm} is the synchronous speed of the machine in mechanical radians/s;
ω_{me} is the speed of the machine in electrical radians/s $= \frac{p}{2}\,.\,\omega_m$
$\omega = 2\pi \times \text{supply frequency} = 2\pi f$;
p is the number of poles on the machine.

$$s = \frac{n_{sm} - n_m}{n_{sm}}$$

Multiply numerator and denominator by $\frac{2\pi}{60}$

$$s = \frac{\omega_{sm} - \omega_m}{\omega_{sm}}$$

Multiply numerator and denominator by $\frac{p}{2}$

$$s = \frac{\omega - \omega_{me}}{\omega}$$

6.2 Three-phase induction machines

6.2.1 *Construction*

A three-phase induction machine carries a three-phase distributed winding on its stator while the rotor may be one of two types as follows.

(a) *Wound rotor machines*

The rotor of a wound rotor machine carries a three-phase distributed winding with the same number of poles as the stator. This winding is usually connected in star with ends of the winding brought out to three slip rings as suggested by Fig. 6.2, enabling external resistance to be added to the rotor for control purposes, particularly during starting.

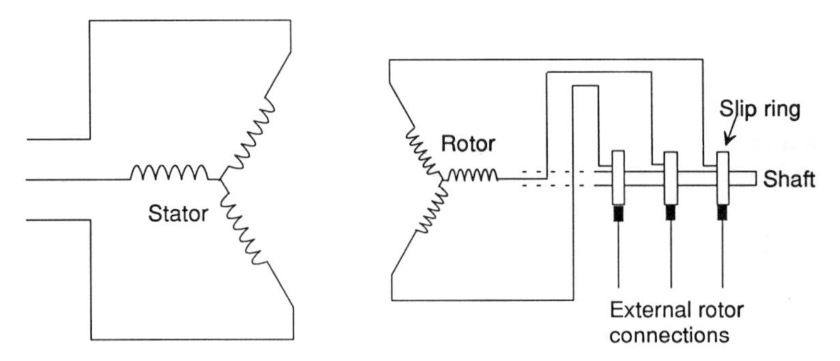

Fig. 6.2 Connections for a wound rotor induction motor.

(b) *Squirrel-cage machines*

The rotor of a squirrel-cage machine carries a winding consisting of a series of bars in the rotor slots which are shorted by end rings at each end of the rotor as in Fig. 6.3. The rotor conductors are often formed by casting once the rotor core has been assembled, a much cheaper form of construction than the wound rotor. In use, the squirrel cage adopts the current pattern and pole distribution of the stator enabling a basic rotor to be used for machines with differing pole numbers. However, for analysis purposes a cage rotor may be treated as a symmetrical, short circuited star-connected three-phase winding.

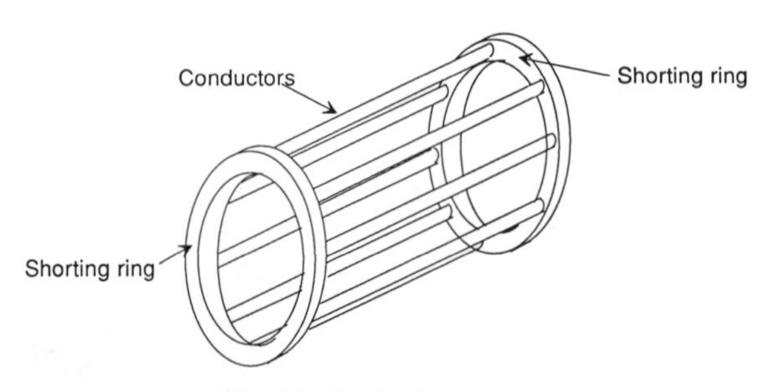

Fig. 6.3 Squirrel cage rotor.

6.2.2 *Equivalent circuit*

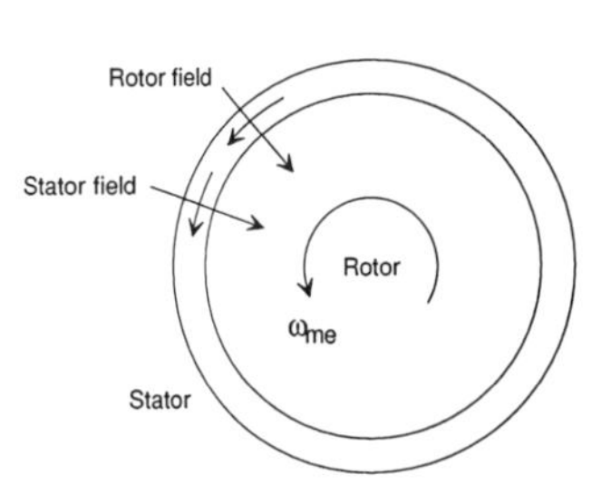

Fig. 6.4 Induction machine stator and rotor fields.

Consider the situation of Fig. 6.4 for an induction machine in which the rotor is moving at an angular velocity of ω_{me}. The relative angular velocity of the stator field with respect to the rotor (ω_r) is then:

$$\omega_r = \omega - \omega_{me} \tag{6.4}$$

where ω is the angular velocity of the stator field in electrical radians/s and ω_{me} is the angular velocity of the rotor in electrical radians/s.

Comment

$$\omega = 2\pi f = \frac{p}{2}\,\omega_{sm}$$

and

$$\omega_{me} = \frac{p}{2}\,\omega_m$$

where p is the number of poles, f is the supply frequency and ω_{sm} is the synchronous speed in mechanical radians per second.

The frequency of the currents induced into the rotor of the machine (f_r) is determined by the difference between the angular velocityof the stator field and the angular velocity of the rotor when, referring to equation (6.4):

$$f_r = \frac{\omega_r}{2\,.\,\pi} \tag{6.5}$$

Referring to the expression for slip of equation (6.2) it can be seen that

$$\omega_r = s\,.\,\omega \tag{6.6}$$

Hence the frequency of the currents induced in the rotor is:

$$f_r = s\,.\,f \tag{6.7}$$

Referring to the rotor, the rotor currents would result in the production of a rotating magnetic field rotating at an angular velocity **relative to the rotor** of ω_r such that:

$$\omega_r = \omega - \omega_{me} \tag{6.8}$$

Hence the rotor conductors 'see' both the stator field and the rotor field as having the same angular velocity of ω_r.

The stator however 'sees' a field due to the rotor having an angular velocity made up of two components, the angular velocity of the rotor (ω_{me}) and the angular velocity of the rotor field relative to the rotor (ω_r). The combination of these two velocities give a solution of:

$$\omega_r + \omega_{me} = \omega \tag{6.9}$$

Thus both the stator and rotor fields have angular velocities **relative to the stator** of ω.

This means that the relationship between the stator and rotor fields is the same whatever reference framework, either stator or rotor, is being used.

Comment

This solution is also independent of the number of poles on the machine.

For ampere turns balance between the rotor and stator:

$$N_s . I_s = N_r . I_r \tag{6.10}$$

Assuming that the resultant airgap flux (made up of the combination of the stator and rotor fluxes) seen by the stator has the form

$$\boldsymbol{\Phi}_s = \hat{\boldsymbol{\Phi}} . \sin \omega t \tag{6.11}$$

The associated stator voltage is then:

$$e_s = N_s . \frac{d\boldsymbol{\Phi}_s}{dt} = \hat{\boldsymbol{\Phi}} . N_s . \omega . \cos \omega t \tag{6.12}$$

which corresponds to an r.m.s. voltage of:

$$E_s = \frac{\hat{\boldsymbol{\Phi}} . N_s . \omega}{\sqrt{2}} \tag{6.13}$$

Relative to the rotor the airgap flux has the same magnitude ($\hat{\boldsymbol{\Phi}}$) but is at rotor frequency hence:

$$\boldsymbol{\Phi}_r = \hat{\boldsymbol{\Phi}} . \sin(s\omega t) \tag{6.14}$$

and the associated rotor voltage

$$e_r = N_r . \frac{d\boldsymbol{\Phi}_r}{dt} = \hat{\boldsymbol{\Phi}} . N_r . s . \omega . \cos(s\omega t) \tag{6.15}$$

which corresponds to an r.m.s. voltage of:

$$E_r = s . \frac{\hat{\boldsymbol{\Phi}} . N_r . \omega}{\sqrt{2}} \tag{6.16}$$

Thus:

$$\frac{E_s}{E_r} = \frac{1}{s} . \frac{N_s}{N_r} \tag{6.17}$$

Knowledge of the form of the transformer equivalent circuit enables the per-phase equivalent circuit of Fig. 6.5 to be produced in which the rotor and the stator are taken

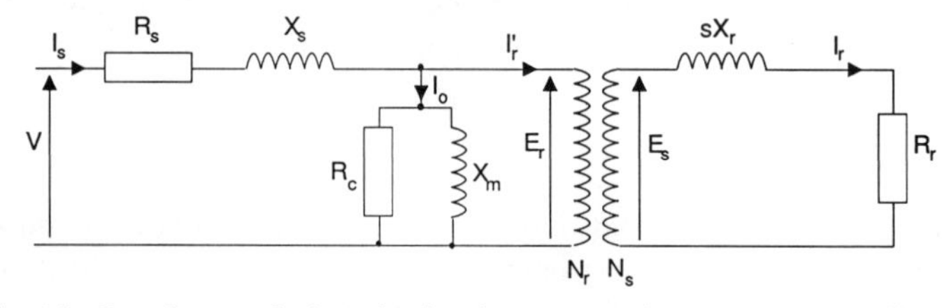

Fig. 6.5 Per-phase equivalent circuit using stator and rotor parameters directly.

to be at different frequencies, as indicated by the incorporation of slip in the expression for rotor reactance.

Comment

See Chapter 3 for the discussion of the transformer.

Rotor impedance may now be transferred to the stator using the relationships of equations (6.10) and (6.15) when, referring to Fig. 6.5:

$$\boldsymbol{Z}_{\rm rs}=\frac{\boldsymbol{E}_{\rm s}}{\boldsymbol{I}_{\rm s}}=\frac{1}{s}\cdot\frac{N_{\rm s}}{N_{\rm r}}\cdot\boldsymbol{E}_{\rm r}\cdot\frac{N_{\rm s}}{N_{\rm r}}\cdot\frac{1}{\boldsymbol{I}_{\rm r}}=\frac{1}{s}\cdot\left(\frac{N_{\rm s}}{N_{\rm r}}\right)^2\cdot\frac{\boldsymbol{E}_{\rm r}}{\boldsymbol{I}_{\rm r}}=\frac{1}{s}\cdot\left(\frac{N_{\rm s}}{N_{\rm r}}\right)^2.\boldsymbol{Z}_{\rm r}$$

$$=\frac{1}{s}\cdot\left(\frac{N_{\rm s}}{N_{\rm r}}\right)^2.(R_{\rm r}+{\rm j}.s.X_{\rm r})=\frac{R_{\rm r}'}{s}+{\rm j}X_{\rm r}'=\frac{1}{s}.\boldsymbol{Z}_{\rm r}' \qquad (6.18)$$

This gives the per-phase equivalent circuit referred to the stator of Fig. 6.6. The stator and magnetizing branch components may be reduced to an equivalent Thevenin circuit form by separating off the rotor at the airgap line of Fig. 6.6 to give the Thevenin equivalent circuit of Fig. 6.7 in which:

$$V_{\rm t}=V_{\rm p}\cdot\left|\frac{\boldsymbol{Z}_{\rm m}}{(\boldsymbol{Z}_{\rm m}+\boldsymbol{Z}_{\rm s})}\right|$$

and

$$Z=\frac{\boldsymbol{Z}_{\rm m}.\boldsymbol{Z}_{\rm s}}{(\boldsymbol{Z}_{\rm m}+\boldsymbol{Z}_{\rm s})}=R+{\rm j}X$$

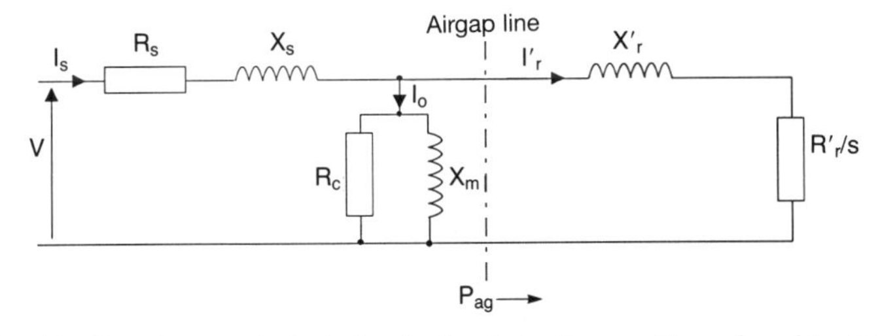

Fig. 6.6 Per-phase equivalent circuit of an induction machine referred to stator.

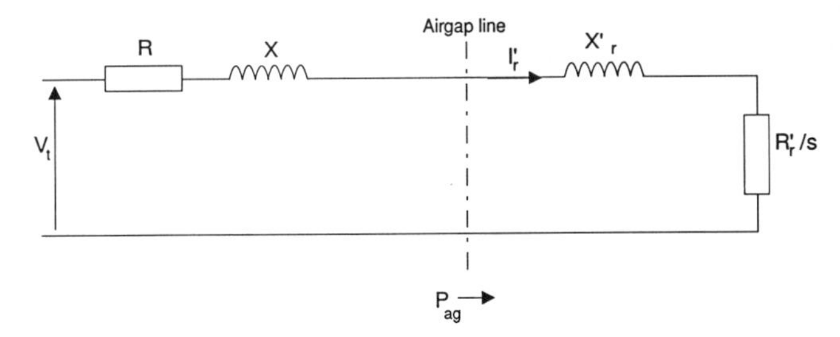

Fig. 6.7 Thevenin per-phase equivalent circuit of an induction machine.

where V_p is the phase voltage and V_t the Thevenin voltage, with

$$\boldsymbol{Z}_m = \frac{R_c \,.\, jX_m}{(R_c + jX_m)}$$

and

$$\boldsymbol{Z}_s = R_s + j \,.\, X_s$$

The solution of this circuit then gives the referred rotor current I_r' directly as:

$$I_r' = \frac{V_t}{\left[\left(\frac{R_r'}{s} + R\right)^2 + (X_r' + X)^2\right]^{\frac{1}{2}}} \tag{6.19}$$

6.2.3 *Torque equation*

The mechanical power (P_m) developed at the shaft of the induction machine is expressed in terms of the mechanical torque (T_m) and the speed (ω_m), when:

$$P_m = T_m \,.\, \omega_m \tag{6.20}$$

The power transferred to the rotor per phase is the airgap power per phase P_{ag} and, in terms of the equivalent circuit, this is the power dissipated in the resistance R_r'/s. Thus:

$$P_{ag} = I_r'^2 \,.\, \frac{R_r'}{s} \tag{6.21}$$

The rotor copper loss per phase (P_r) is determined by the actual rotor resistance (R_r) and rotor current (I_r) as:

$$P_r = I_r^2 \,.\, R_r = I_r'^2 \,.\, \left(\frac{N_s}{N_r}\right)^2 \,.\, R_r = I_r'^2 \,.\, R_r' \tag{6.22}$$

The mechanical power output per phase (P_{mp}) is:

$$P_{mp} = \text{Airgap power per phase} - \text{Rotor copper loss per phase} = P_{ag} - P_r$$

$$= I_r'^2 \,.\, R_r' \,.\, \left(\frac{1}{s} - 1\right) = P_{ag}(1 - s) \tag{6.23}$$

Comment

The R_r'/s term for rotor resistance in the equivalent circuit could therefore be replaced by two resistors of value R_r' and $R_r'(1-s)/s$, respectively. The first of these (R_r') would represent the rotor copper loss and the second [$R_r'(1-s)/s$] the mechanical power developed.

Now, from the definition of slip:

$$\omega_m = \omega_{sm}(1 - s) \tag{6.24}$$

where ω_{sm} is the synchronous speed in mechanical radians per second.

Combining equations (6.20), (6.23) and (6.24) for an m phase machine:

$$T_m = \frac{m \,.\, P_{ag}}{\omega_{sm}} \tag{6.25}$$

Using equation (6.19) for I_r' gives:

$$P_{ag} = \frac{V_t^2}{\left[\left(\frac{R_r'}{s} + R\right)^2 + (X_r' + X)^2\right]} \cdot \frac{R_r'}{s} \tag{6.26}$$

when

$$T_m = \frac{m}{\omega_{sm}} \cdot \frac{V_t^2}{\left[\left(\frac{R_r'}{s} + R\right)^2 + (X_r' + X)^2\right]} \cdot \frac{R_r'}{s} \tag{6.27}$$

In terms of the supply frequency, putting $\omega_{sm} = \dfrac{2\omega}{p}$.

$$T_m = \frac{m \,.\, p}{2 \,.\, \omega} \cdot \frac{V_t^2}{\left[\left(\frac{R_r'}{s} + R\right)^2 + (X_r' + X)^2\right]} \cdot \frac{R_r'}{s} \tag{6.28}$$

This gives a torque/slip (torque/speed) characteristic for the three-phase induction machine of the form of Fig. 6.8.

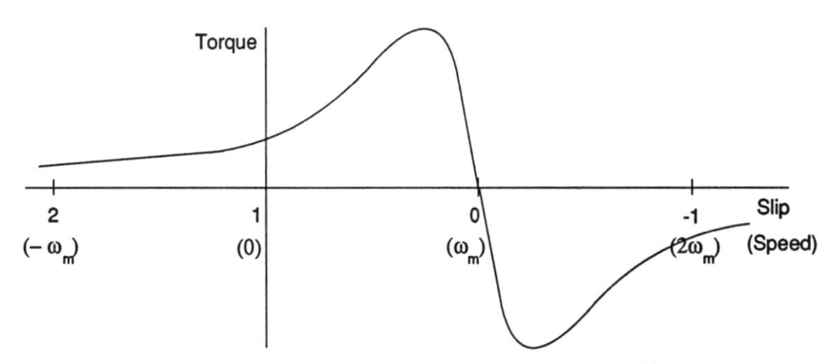

Fig. 6.8 Torque/slip characteristic of an induction machine.

(a) *Power flow in a three-phase induction machine*

Equation (6.28) defines the gross mechanical torque produced, and hence the gross mechanical power developed, by a three-phase induction machine. The useful mechanical torque will take account of the rotational losses in the machine when:

$$\text{Useful mechanical power} = \text{Gross mechanical power developed} - \text{Rotational losses} \tag{6.29}$$

The power flow diagram for the three-phase induction machine when motoring is then as shown in Fig. 6.9.

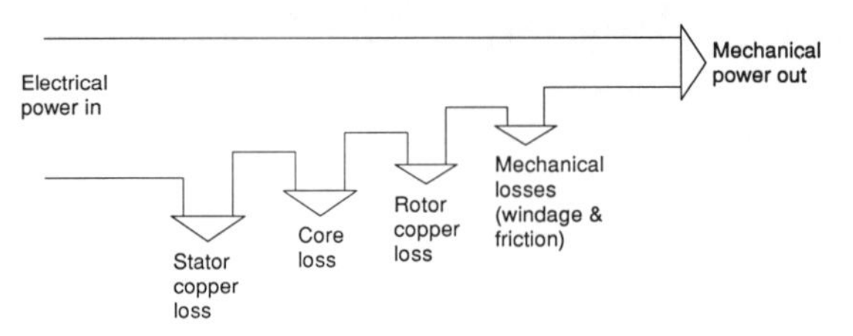

Fig. 6.9 Induction motor energy flow diagram (losses not to scale).

Comment

The rotational losses are made up of air resistance or windage and friction.

Example 6.1 Three-phase induction machine
A 50 Hz, three-phase, 415 volt (line), four-pole induction slip-ring induction machine when running at a speed of 1415 rev/min has windage and friction losses of 520 watts. The machine parameters per phase (assuming star connection) are:

$R_r = 0.08\ \Omega$

$X_r = 0.15j\ \Omega$

$R_s = 0.12\ \Omega$

$X_s = 0.3j\ \Omega$

$R_c = 380\ \Omega$ Referred to stator

$X_m = 150j\ \Omega$ Referred to stator

$\frac{N_s}{N_r} = 1.5$

Estimate the mechanical torque developed, the useful mechanical power and the efficiency of the machine.

Calculate the Thevenin values for the equivalent circuit.

$$Z_m = \frac{380 \times 150j}{380 + 150j} = 139.52\angle 68.46^\circ = 51.23 \text{ [illegible] } 9.78j\ \Omega$$

and

$$Z_s = 0.12 + 0.3j = 0.323\angle 68.2 \text{ [illegible]}$$

when

$$Z_s + Z_m = 51.35 + 130.08j = 139 \text{ [illegible]}$$

and

$$Z_t = \frac{139.52\angle 68.46^\circ \times 0.323\angle 68.2^\circ}{139.85\angle 68.46^\circ} = \text{ [illegible] } + 0.3j\ \Omega = R + jX\ \Omega$$

and

$$V_t = V_{phase} \cdot \left(\frac{Z_m}{Z_s + Z_m} \right) = \frac{415}{\sqrt{3}} \cdot \left| \frac{139.52\angle 68.46^\circ}{139.85\angle 68.46^\circ} \right| = 239 \text{ V}$$

Slip

$$s = \frac{1500 - 1415}{1500} = 0.05667$$

Referring rotor impedance values to stator.

$$R'_r = 0.08 \times 1.5^2 = 0.18\ \Omega$$

when

$$\frac{R'_r}{s} = 3.18$$

and

$$X'_r = 0.15 \times 1.5^2 = 0.338\ \Omega$$

Substituting in equation (6.28)

$$T_m = \frac{3 \times 4}{2 \times 314.16} \cdot \left[\frac{239^2}{(3.18 + 0.12)^2 + (0.338 + 0.3)^2} \right] \times 3.18 = 307.1 \text{ N m}$$

The useful mechanical power is then:

$$P_0 = 307.1 \times 2\pi \times \frac{1415}{60} - 520 = 44\ 990 \text{ W}$$

The input power and efficiency may be estimated using the Thevenin equivalent circuit when, with V_t as reference:

$$I'_r = \frac{239}{3.3 + 0.638\text{j}} = 71.11\angle -10.94^\circ \text{ A}$$

Input power is then

$$P_{in} = 3 \times 239 \times 71.11 \times \cos(10.94^\circ) = 50\ 060 \text{ W}$$

Hence efficiency:

$$\eta = 100 \times \frac{44\ 990}{50\ 060} = 89.87\%$$

Or, more accurately, voltage across magnetizing branch:

$$V_m = I'_r \left(\frac{R'_r}{s} + \text{j}Xr' \right) = 227.4\angle -4.86^\circ \text{ V}$$

Hence current in R_c

$$I_c = \frac{V_m}{380} = 0.598\angle -4.86^\circ \text{ A}$$

and in X_m

$$\boldsymbol{I}_m = \frac{\boldsymbol{V}_m}{150j} = 1.52\angle{-94.86°}\ \text{A}$$

Hence stator current

$$\boldsymbol{I}_s = \boldsymbol{I}'_r + \boldsymbol{I}_c + \boldsymbol{I}_m = 71.88\angle{-12.1°}\ \text{A}$$

Then

$$P_{ag} = I_r'^2 \cdot \frac{R'_r}{s} = 16\ 080\ \text{W/phase}$$

Power lost in R_c

$$P_c = I_c^2 \cdot R_c = 135.9\ \text{W/phase}$$

Power lost in stator

$$P_s = I_s^2 \cdot R_s = 620\ \text{W/phase}$$

Total input power is then

$$P_{in} = 3 \times (16\ 080 + 135.9 + 620) = 50\ 508\ \text{W}$$

Hence efficiency

$$\eta = 100 \times \frac{44\ 990}{50\ 508} = 89.07\%$$

(b) *Maximum torque*

Referring to equation (6.28), maximum torque occurs when:

$$\frac{dP_{ag}}{ds} = 0 \tag{6.30}$$

Solving equation (6.30), the slip at which maximum torque occurs is then:

$$s_{max} = \frac{R'_r}{[R^2 + (X + X'_r)^2]^{\frac{1}{2}}} \tag{6.31}$$

Substituting for s_{max} in equation (6.28), the maximum torque (T_{max}) is then:

$$T_{max} = \frac{m \cdot p}{4 \cdot \omega} \cdot \frac{V_t^2}{R + [R^2 + (X + X'_r)^2]^{\frac{1}{2}}} \tag{6.32}$$

Inspection of equations (6.31) and (6.32) shows that for fixed stator and magnetizing branch values, the slip at which maximum torque occurs is determined by the value of the rotor resistance R'_r with increasing values of R'_r causing s_{max} to increase. However, the value of maximum torque developed by the machine is independent of the rotor resistance. Figure 6.10 illustrates the effect of varying rotor resistance on the performance of the machine.

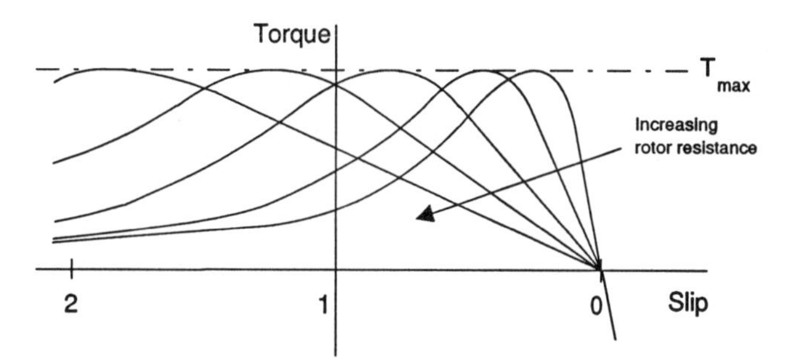

Fig. 6.10 Effect of varying rotor resistance on induction motor torque characteristic.

Example 6.2 Maximum torque
Calculate the maximum torque and the slip at which it occurs for the machine of Example 6.1. What value of resistance must be added to each phase of the rotor to ensure that maximum torque occurs at zero speed.

Using equation (6.31)

$$s_{\max} = \frac{0.113}{(0.239^2 + 0.803^2)^{\frac{1}{2}}} = 0.1363$$

which corresponds to a speed of 1295.5 rev/min.

The maximum torque is then obtained by using equation (6.32) when:

$$T_{\max} = \frac{3 \times 4}{4 \times 314.16} \cdot \frac{238.6^2}{0.239 + (0.239^2 + 0.803^2)^{\frac{1}{2}}} = 505.5 \text{ N m}$$

For maximum torque at starting

$$R_r' = (0.239^2 + 0.803^2)^{\frac{1}{2}} = 0.838\ \Omega$$

Added rotor resistance per phase is then

$$R_{\text{radd}} = \frac{0.838}{1.5^2} - 0.05 = 0.322\ \Omega/\text{phase}$$

6.2.4 *Three-phase induction machine operation*

From Fig. 6.11 it can be seen that the induction machine has three operation regimes; namely **motoring**, **generating** and **braking**, depending on the value of slip.

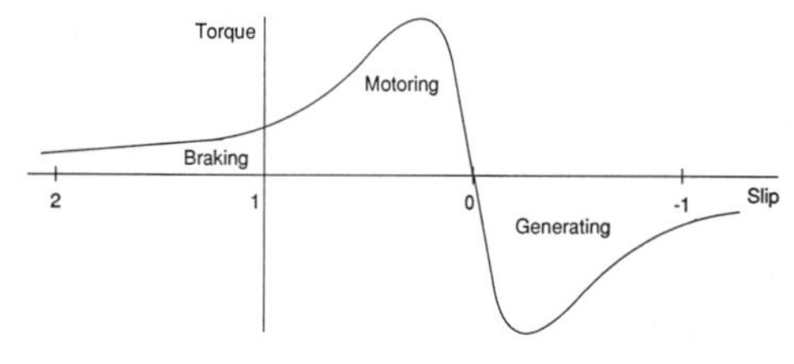

Fig. 6.11 Induction machine operating regimes.

(a) *Motoring*

When motoring the slip is in the region 1 to zero and the rotor is moving in the same direction as the stator field. Typical operation is at a point, such as that shown in Fig. 6.12, on the downslope of the characteristic near to synchronous speed.

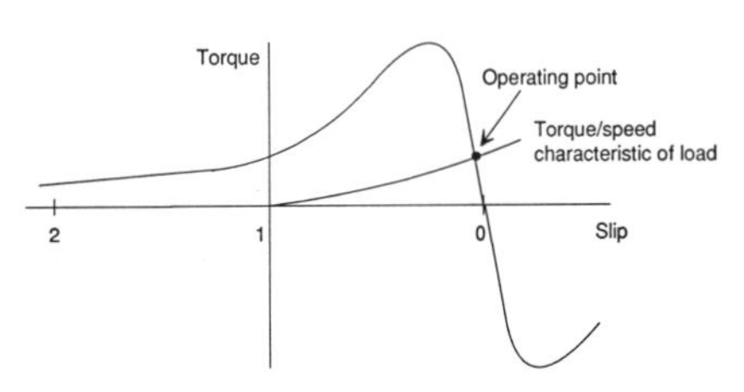

Fig. 6.12 Induction machine – operation in motoring mode.

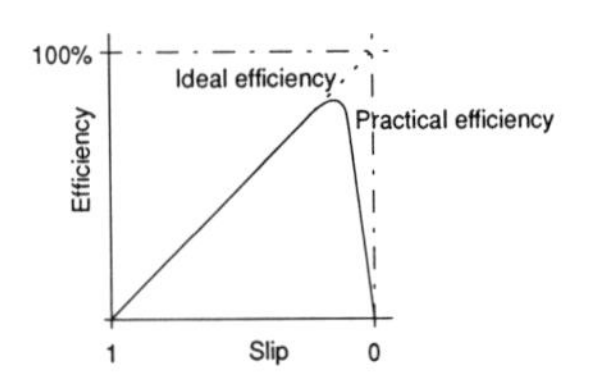

Fig. 6.13 Ideal and practical efficiencies of an induction motor.

Ignoring stator and rotational losses, the efficiency of an induction motor can be approximated as:

$$\text{Efficiency} = \eta = \frac{\text{Mechanical power out}}{\text{Electrical power transferred across airgap}} = \frac{P_{\mathrm{m}}}{P_{\mathrm{ag}}} = (1 - s) \quad (6.33)$$

which has the form shown in Fig. 6.13.

Comment

The result of equation (6.33) ignores all losses in the induction machine and would result in an efficiency of 100% at a slip of zero (synchronous speed). In practice, the efficiency falls off as synchronous speed is approached as shown by the solid line in Fig. 6.13. Maximum efficiencies range from percentage values in the mid-sixties for small machines to the high nineties for large machines.

The starting torque of the machine is found by putting $s = 1$ in equation (6.28) and is associated with a starting current which is typically of the order of 7 to 10 times the normal full load current of the machine. In the case of the larger induction motors this may result in a severe dip in the voltage supplied to the machine during starting and various methods are used to modify the starting performance and reduce the starting current.

Star/delta starter

A star–delta starter initially connects the stator of the machine in star so that the full impedance per phase is presented to the phase voltage. When the speed of the machine has reached a particular value the stator is reconnected in delta, reducing the effective per-phase impedance for the equivalent star to one-third of its previous value.

Added rotor resistance

In the case of the wound rotor machine additional resistance may be connected into the rotor via the slip-rings to modify the machine characteristic and reduce the starting current. As the machine accelerates the added resistance is removed to give normal machine operation.

For a squirrel cage machine, the effective rotor resistance on starting can be increased by means of the double cage construction of Fig. 6.14(a). Initially, all the rotor current is confined to the high impedance outer cage, increasing starting torque and reducing the starting current. As the speed increases, the proportion of the total rotor current carried by the low impedance inner cage increases giving the generalized characteristic of Fig. 6.14(b).

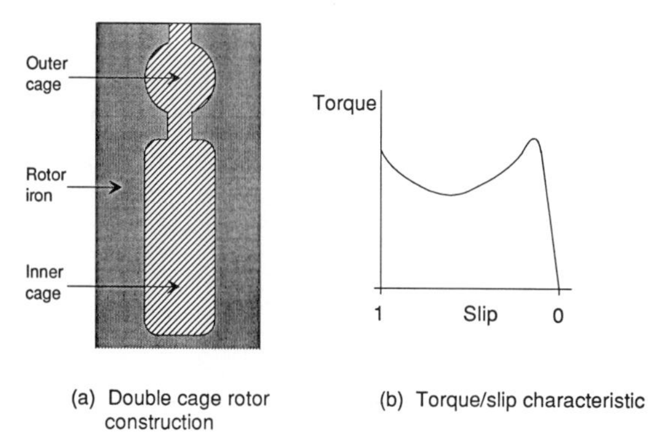

Fig. 6.14 Induction machine with double cage rotor: (a) double cage rotor construction; and (b) torque/slip characteristic.

Soft-start

A soft-start system applies a gradually increasing voltage to the stator of the machine throughout the whole of the acceleration period. The applied voltage and its variation is usually controlled using inverse parallel thyristor pairs as in Fig. 6.15 to limit the stator current during acceleration and run-up.

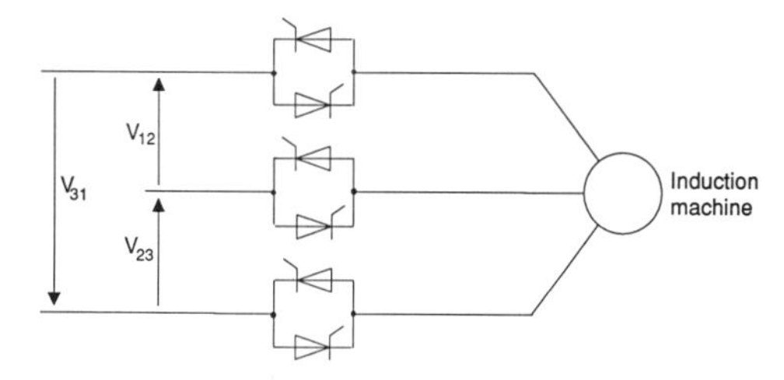

Fig. 6.15 Configuration of a soft starter for an induction motor using inverse thyristor pairs.

(b) *Generating*

When operating in the generating mode the rotor of the machine is travelling at a speed greater than synchronous speed (**super-synchronous operation**) and slip is therefore negative. In terms of the equivalent circuit this means that the R_r'/s term becomes negative in which case it acts as a source of power rather than a sink and energy is returned to the supply.

Typically, an induction generator would be operated connected to a fixed frequency supply to provide the necessary reference.

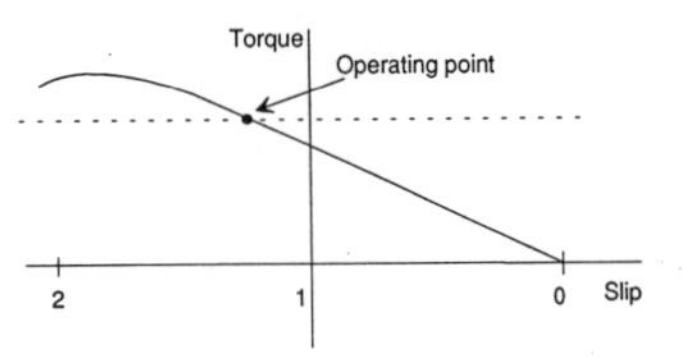

Fig. 6.16 Induction machine – operation in braking mode.

(c) *Braking*

When operating in braking mode the direction of rotation of the rotor is in the opposite direction to the direction of the stator field. Under these conditions slip is greater than 1 and the resulting torque opposes the motion of the rotor.

Operation in braking mode is normally a very severe condition unless controlled by the introduction of additional resistance into the rotor in which case it is possible to operate in a stable condition on the forward slope of the characteristic as in Fig. 6.16.

Example 6.3 Induction generator
The machine of Example 6.1 is connected to a three-phase, 415 volt (line) supply and driven at a speed of 1530 rev/min when used as an induction generator. Determine the power that is returned to the supply under these conditions.

Slip

$$s=\frac{1500-1530}{1500}=-0.02$$

therefore

$$\frac{R_r'}{s}=-5.65$$

when

$$\boldsymbol{Z}=-5.411+0.803\mathrm{j}=5.47\underline{|171.56°}\ \Omega$$

Hence

$$\boldsymbol{I}_r'=\frac{238.6}{5.47}\underline{|-171.56°}=88.11\underline{|-171.56°}\ \mathrm{A}$$

Therefore

$$\text{Generated power}=3\times238.6\times88.11\times\cos(8.44°)=30\ 880\ \mathrm{W}$$

6.2.5 *Induction machine testing*

The values of the equivalent circuit parameters for an induction machine are found by means of the locked rotor and the open circuit or no-load tests.

(a) *Locked rotor test*

In the locked rotor test the rotor is constrained to prevent it rotating and a voltage applied to the stator such as to drive rated current through the stator windings. The applied voltage per phase (V_p), the current (I) and the per-phase power (P) supplied are all recorded when, since slip has a value of 1 at standstill:

$$\boldsymbol{Z}=[(R_s+R_r')^2+(X_s+X_r')^2]^{\frac{1}{2}}=\frac{V_p}{I} \tag{6.34}$$

and

$$P = I^2(R_s + R_r') \qquad (6.35)$$

Hence the values for $(R_s + R_r')$ and $(X_s + X_r')$ can be found. As the value of the stator resistance (R_s) can be found by direct measurement, the value of R_r' can be established.

Comment

The locked rotor test should be compared with the short circuit test for a transformer.

(b) *Open circuit test – wound rotor induction machines*

In the open circuit test on a wound rotor induction machine the rotor windings are open circuited at the slip-rings and rated voltage is then applied to the stator winding. The stator phase voltage (V_p), the stator current (I) and the power supplied are recorded as before.

Under these conditions the effect of the stator impedance can be ignored in which case, referring to Fig. 6.16:

$$P = \frac{V_p^2}{R_c} \qquad (6.36)$$

when

$$I_c = \frac{V_p}{R_c} \qquad (6.37)$$

Hence

$$I_m = (I^2 - I_c^2)^{\frac{1}{2}} \qquad (6.38)$$

when

$$X_m = \frac{V_p}{I_m} \qquad (6.39)$$

Comment

The open circuit test should be compared with the open circuit test of a transformer described in Section 3.5.3.

(c) *No-load test – squirrel cage and wound rotor induction machines*

In the case of a squirrel cage induction machine it is impossible to open circuit the rotor and an alternative means must be found to obtain the values of the magnetizing branch resistance (R_c) and reactance (X_m). If the induction machine is operated with no external applied load the speed of the machine will be very close to synchronous speed and the slip will, therefore be very small. Under these conditions, the R_r'/s term in the equivalent circuit will be large, enabling it to be considered as an effective open circuit for the purpose of the test. By measuring applied voltage, stator current and the power supplied the value of the magnetizing branch components can be calculated using equations (6.36) to (6.39) for the open circuit test on a wound rotor machine.

6.3 Induction machine control

6.3.1 *Armature resistance control*

By adding resistance to the rotor the machine characteristic can be shaped to provide speed control as in Fig. 6.17. The range over which speed can be varied is limited and the efficiency is low at the lower end of the controlled range.

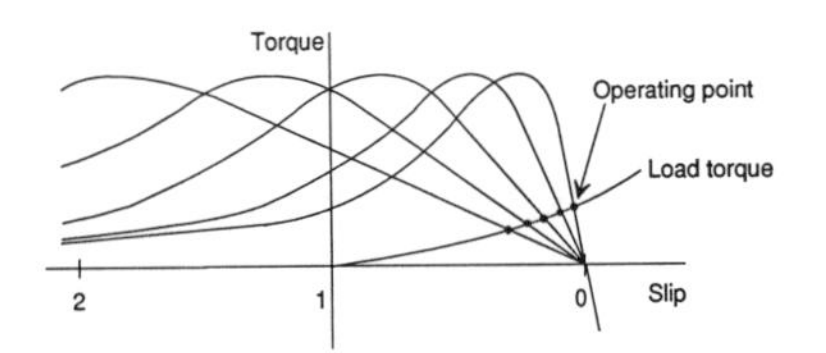

Fig. 6.17 Induction motor speed control using added rotor resistance.

6.3.2 *Pole-amplitude modulation*

Pole-amplitude modulation utilizes a stator winding which can be reconfigured by external switching to produce differing numbers of pole-pairs thus changing the effective synchronous speed at which the machine operates.

6.3.3 *Armature voltage control*

A limited degree of speed control can be achieved by varying the voltage applied to the stator of the induction motor, producing a family of characteristics of the form shown in Fig. 6.18. The range of speed variation that can be achieved using this form of control is limited. It is, however, used as the basis of 'energy saver' controls to reduce the current drawn by the machine when running on no-load.

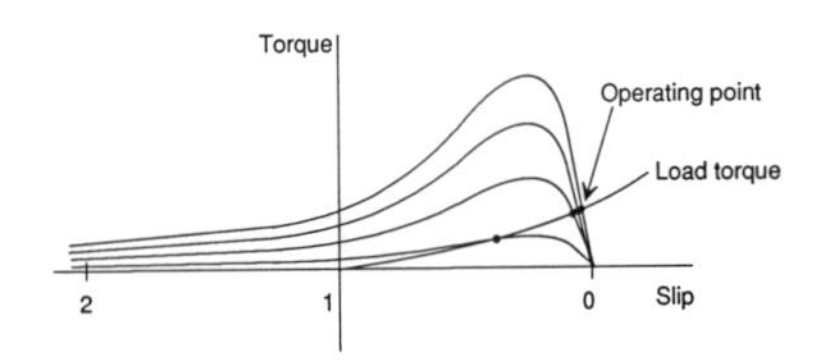

Fig. 6.18 Induction motor speed control by varying supply voltage magnitude.

6.3.4 *Slip energy recovery*

Armature resistance control enables the speed of an induction machine to be controlled by extracting energy from the rotor. If this energy is returned to the supply instead of being dissipated in a resistive load the overall efficiency of the system can be maintained at a reasonable value over the full controlled range of the machine.

Slip energy recovery systems therefore extract energy from the rotor of a wound rotor machine at slip frequency and then return it via a frequency converter to the three-phase supply. A typical configuration for a **static Kramer drive** using a rectifier and a controlled converter is shown in Fig. 6.19.

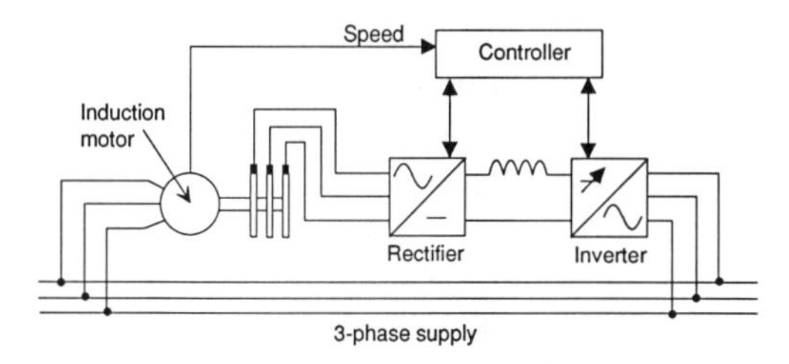

Fig. 6.19 Static Kramer drive.

6.3.5 *Variable frequency control*

Ideally, an induction machine should operate with a constant flux throughout the whole of its speed range. Referring to equation (6.13), this gives a relationship between the stator voltage E_s and the supply frequency (f) for constant flux conditions of the form:

$$\frac{E_s}{f} = \sqrt{2}\pi\hat{\Phi} \, . \, N_s = \text{constant} \tag{6.40}$$

Ignoring stator resistance, equation (6.40) can be approximated in terms of the stator voltage as:

$$\frac{V}{f} = \text{constant} \tag{6.41}$$

Also

$$T_{max} = \frac{m \, . \, p}{4 \, . \, \omega} \cdot \frac{V_t^2}{R + [R^2 + (X + X_r')^2]^{\frac{1}{2}}} \tag{6.42}$$

which is independent of speed resulting in a torque/speed regime for an ideal machine with V/f constant of the form of Fig. 6.20. In practice, for frequencies above the nominal supply frequency the stator voltage would be held constant as the frequency is varied giving the overall torque/speed characteristic for an ideal induction machine of the form of Fig. 6.21.

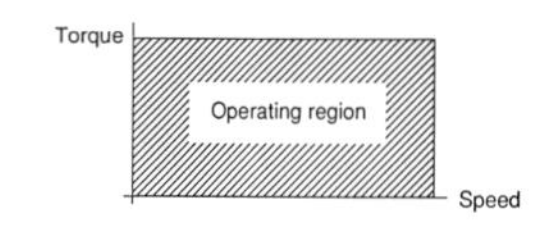

Fig. 6.20 Idealized torque/speed regime for an induction motor operated with V/f constant.

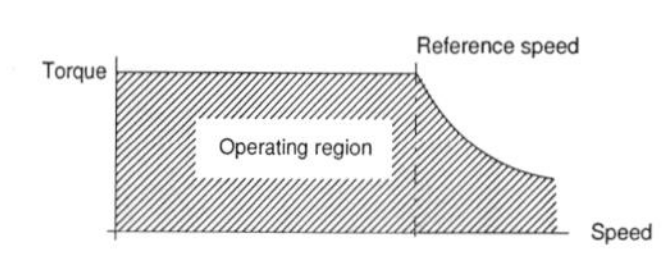

Fig. 6.21 Idealized torque/speed characteristic of an induction motor for operation above and below the reference speed $(=\omega)$.

Comment

Figure 6.21 should be compared with Figs 4.39 and 4.40 for a d.c. machine operating with armature voltage and field control.

In practice, at low speeds the effect of the stator resistance is to cause a reduction in the effective voltage which results in a decrease in the maximum torque capability at low speeds, as illustrated by Fig. 6.22.

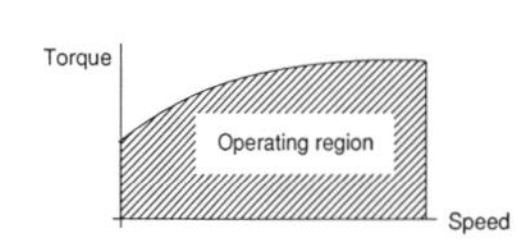

Fig. 6.22 Effect of rotor resistance on low speed torque of an induction motor operated with V/f constant.

Typically, variable frequency control of an induction machine would use a voltage sourced pulse width modulated (PWM) inverter as in Fig. 6.23. To overcome the effect of rotor resistance at low speed referred to in Fig. 6.22, the inverter voltage would normally be increased (boosted) at low speeds as in Fig. 6.24. In practice, a single boost setting is used, though the possibility of continuously modulating the boost across the full speed range of the machine has been investigated.

Figure 6.25 shows a typical control system for a voltage sourced PWM inverter drive. The speed demand signal (ω_{ref}) is compared with the output of the tachogenerator (ω_{tach}) to produce an error signal (ω_{err}). This error signal is then supplied to a regulator which generates a slip speed (ω_{slip}) limited to a maximum value which corresponds to the maximum operating torque of the machine to prevent stalling. The slip speed is then combined with the actual speed of the machine (ω_{tach}) to determine the required synchronous speed (ω_s) and hence the frequency of the inverter. The reference synchronous speed setting is then used to generate the control signal for the operation of the inverter.

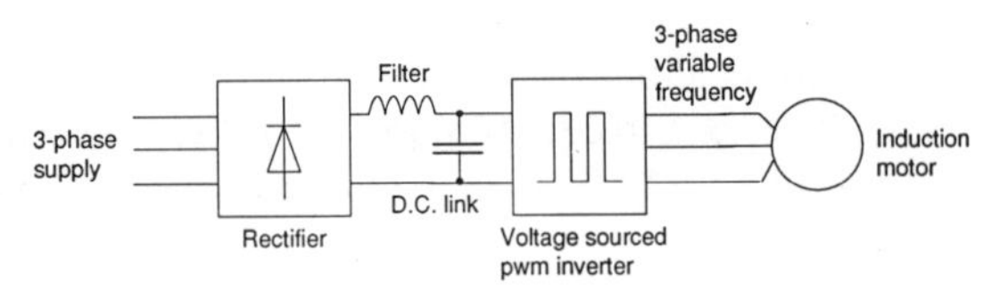

Fig. 6.23 Variable frequency control of an induction motor using a voltage sourced PWM inverter.

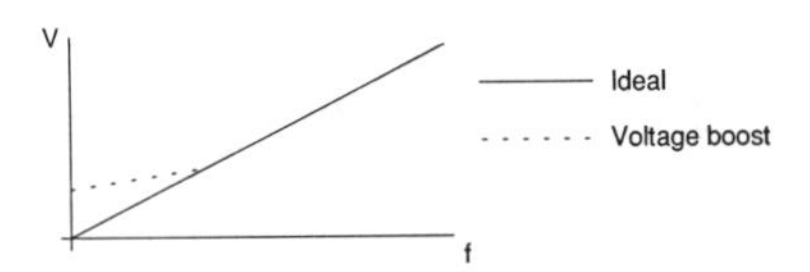

Fig. 6.24 Voltage to frequency relationship for a PWM inverter showing the introduction of voltage boost at low frequency.

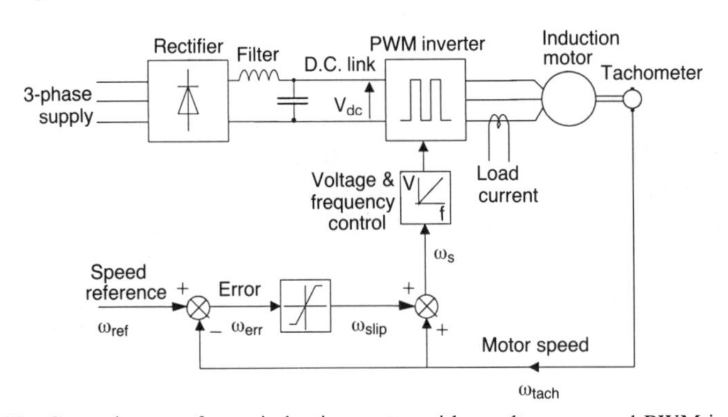

Fig. 6.25 Control system for an induction motor with a voltage sourced PWM inverter.

Example 6.4 Speed control

For the machine of Example 6.1 and taking 50 Hz operation as reference, estimate the maximum torque that will be produced assuming V/f = constant at supply frequencies of 25 Hz and 10 Hz. To what value must the 10 Hz voltage be increased in order to develop the same maximum torque as at 50 Hz?

Using the Thevenin equivalent circuit at 25 Hz

$$Z_{t25} = 0.239 + 0.199\text{j}\ \Omega$$

$$V_t = 119.3\ \text{V}$$

and

$$X'_r = 0.203\ \Omega$$

Hence

$$T_{max;25} = 388.7\ \text{N m}$$

Using the Thevenin equivalent circuit at 10 Hz

$$Z_{t;10} = 0.239 + 0.08\text{j}\ \Omega$$

$$V_t = 47.72\ \text{V}$$

and

$$X'_r = 0.081\ \Omega$$

$$T_{max;10} = 206.3\text{N m}$$

Hence, to restore a maximum torque equal to that obtained at 50 Hz:

$$V_t = \left(\frac{T_{max;50} \times 20\pi \times \{0.239 + [0.239^2 + (0.08 + 0.081)^2]^{\frac{1}{2}}\}}{3}\right)^{\frac{1}{2}} = 74.57\ \text{V}$$

5.3.6 *Braking*

(a) *d.c. dynamic braking*

For d.c. dynamic braking the a.c. supply to the motor is disconnected and a d.c. supply connected as in Fig. 6.26(a). This produces a stationary magnetic field in the airgap and results in a braking torque characteristic of the form of Fig. 6.26(b).

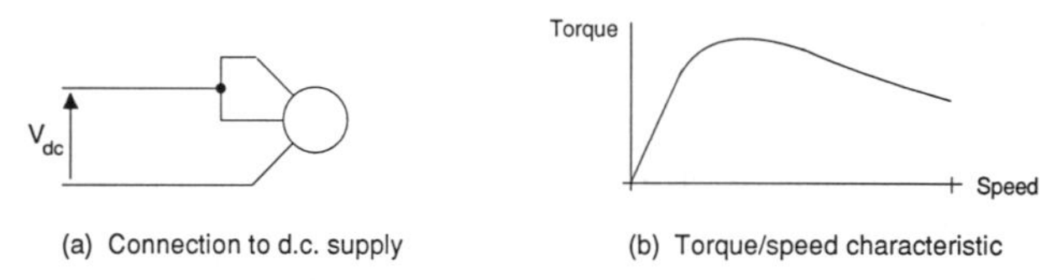

Fig. 6.26 d.c. dynamic braking of an induction motor: (a) connection to d.c. supply; and (b) torque/speed characteristic.

(b) *Regenerative braking*

If for an induction machine operating from a variable frequency inverter, the output frequency of the inverter is reduced so that the machine is running above synchronous speed it operates in generating mode and returns power to the d.c. link. Through the provision of a fully controlled converter operating in inverting mode as suggested by Fig. 6.27, energy can be returned to the a.c. system.

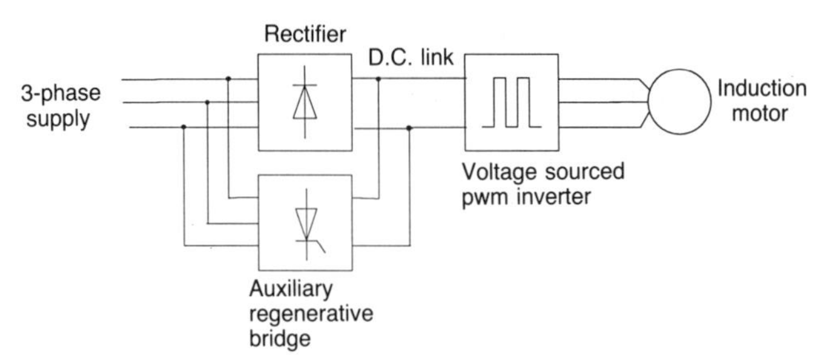

Fig. 6.27 Regenerative braking of an induction motor by means of an auxiliary bridge.

Comment

See Chapter 2 for a discussion of a fully controlled converter and operation in inverting mode.

(c) *Resistive braking*

As an alternative to providing a regenerating inverter, a braking resistor together with a controlled switch can be used as shown in Fig. 6.28 to absorb the energy returned to the d.c. link by the motor when regenerating.

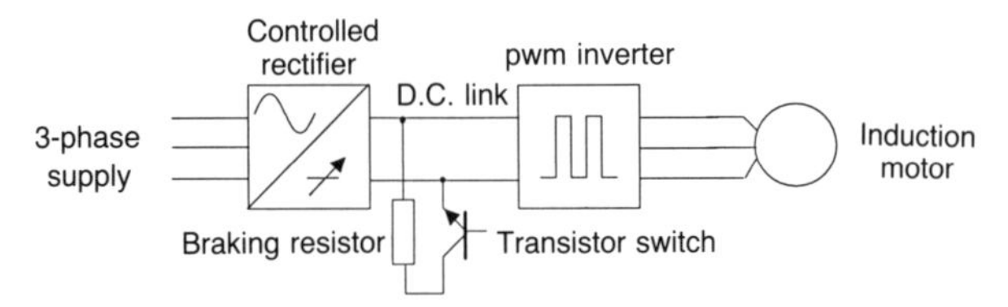

Fig. 6.28 Induction motor: resistive braking.

6.4 Single-phase induction machines

In Chapter 3 it was seen that the pulsating field produced by a single coil when supplied with an a.c. current could be represented by a pair of contra-rotating fields of equal amplitude. Consider the arrangement of Fig. 6.29 with two windings – the main winding and the auxiliary or starting winding – aligned at some angle (φ) and carrying

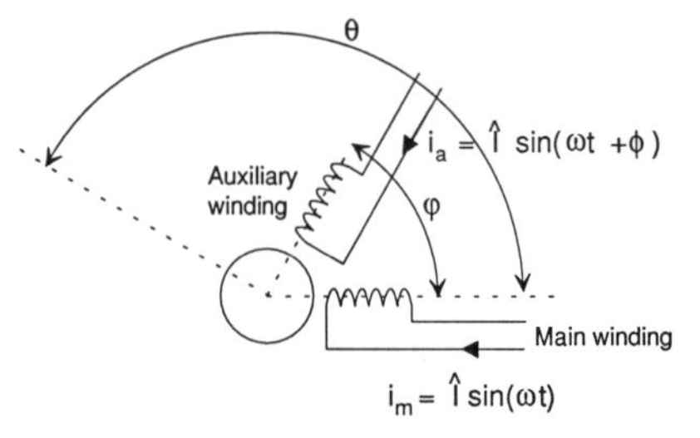

Fig. 6.29 Single-phase induction machine – main and auxiliary windings.

currents of the same frequency but differing in phase by some angle (ϕ). The fields produced at some physical angle (θ) will then be:

Main winding
$$\boldsymbol{F}_{\theta \mathrm{m}}=\hat{F}_{\mathrm{m}} \,.\, \cos \theta \,.\, \cos \omega t$$
$$=\frac{\hat{F}_{\mathrm{m}}}{2} \,.\, [\cos(\theta-\omega t)+\cos(\theta+\omega t)] \tag{6.43}$$

Auxiliary winding
$$\boldsymbol{F}_{\theta \mathrm{a}}=\hat{F}_{\mathrm{a}} \,.\, \cos(\theta-\varphi) \,.\, \cos(\omega t-\phi)$$
$$=\frac{\hat{F}_{\mathrm{a}}}{2} \,.\, [\cos(\theta-\omega t+\phi-\varphi)+\cos(\theta+\omega t-\phi-\varphi)] \tag{6.44}$$

The resultant m.m.f. at angle θ is then:

$$\boldsymbol{F}_{\theta}=\hat{F}_{\mathrm{f}} \,.\, \cos(\theta-\omega t+\sigma_1)+\hat{F}_{\mathrm{b}} \,.\, \cos(\theta+\omega t+\sigma_2) \tag{6.45}$$

in which

$$\hat{F}_{\mathrm{f}}=\frac{1}{2}\{[\hat{F}_{\mathrm{m}}+\hat{F}_{\mathrm{a}} \,.\, \cos(\phi-\varphi)]^2+\hat{F}_{\mathrm{a}}^2 \,.\, \sin^2(\phi-\varphi)\}^{\frac{1}{2}} \tag{6.46}$$

representing the forward travelling field and

$$\hat{F}_{\mathrm{b}}=\frac{1}{2}\{[\hat{F}_{\mathrm{m}}+\hat{F}_{\mathrm{a}} \,.\, \cos(\phi+\varphi)]^2+\hat{F}_{\mathrm{a}}^2 \,.\, \sin^2(\phi+\varphi)\}^{\frac{1}{2}} \tag{6.47}$$

representing the backward travelling field.

Comment

For two windings physically aligned at 90° ($\pi/2$) with currents which are equal in amplitude and out of phase by 90° ($\pi/2$) then:

$$\hat{F}_{\mathrm{f}}=\frac{\hat{F}_{\mathrm{m}}}{2}$$

and

$$\hat{F}_{\mathrm{b}}=0$$

This is the result for a symmetrical two-phase winding with balanced two-phase current and gives a forward rotating field and hence a forward torque only.

By suitable choice of coil currents and the angles φ and ϕ, the magnitude of the forward rotating field can be made to be much greater than that of the backward rotating field with the result that the forward torque dominates and the machine will accelerate in that direction to a stable speed at which point:

- The induced currents in the rotor due to the forward rotating field have a frequency of $s\omega$.
- The induced currents in the rotor due to the backward rotating field have a frequency of $(2-s)\omega$.

Once the machine is running in the forward direction the auxiliary winding can be disconnected in which case the expression for $\boldsymbol{F}_\theta$ becomes:

$$\boldsymbol{F}_\theta = \frac{\hat{F}_{\mathrm{m}}}{2} \cdot [\cos(\theta - \omega t) + \cos(\theta + \omega t)] \tag{6.48}$$

The equivalent circuit for the single-phase induction motor then has the form shown in Fig. 6.30 in which, as only half the total stator current contributes to each of the forward and backward components of the field, the values of the rotor impedances and of the magnetizing reactance used in each part of the circuit are halved. The power in the $R_{\mathrm{r}}'/2s$ resistor is then associated with the forward torque developed by the machine and that in the $R_{\mathrm{r}}'/[2(2-s)]$ resistor with the backward torque.

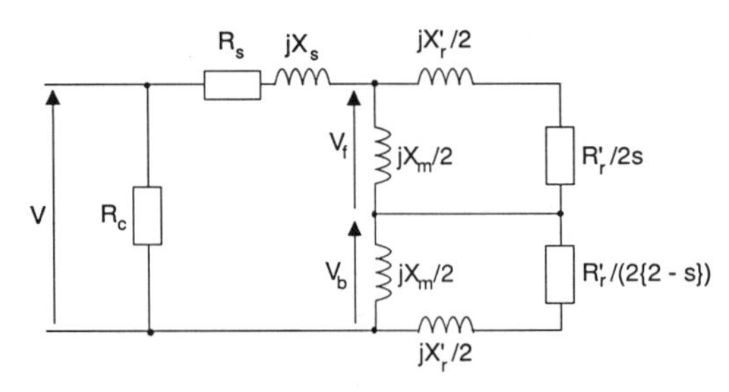

Fig. 6.30 Equivalent circuit of a single-phase induction machine.

Comment

Note that the resistor R_{c} representing the core loss has been included across the terminals of the machine and not in parallel with the magnetizing impedance as before.

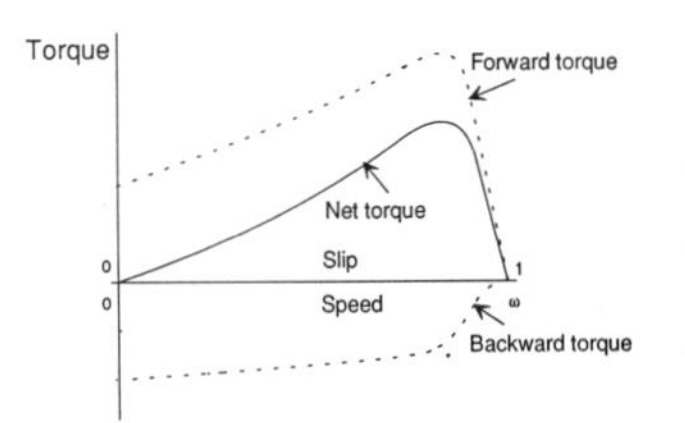

Fig. 6.31 Torque characteristic of a single-phase machine.

The torque/slip (torque/speed) characteristic for the single-phase machine is then as shown in Fig. 6.31.

Example 6.5 Single-phase induction machine
A four-pole, 240 volt single-phase induction machine has the following equivalent circuit parameters:

$R_{\mathrm{s}} = 8\ \Omega$

$X_{\mathrm{s}} = 11.4\ \Omega$

$R_{\mathrm{r}}' = 9.6\ \Omega$

$X_{\mathrm{r}}' = 11.8\ \Omega$

$R_{\mathrm{c}} = 2400\ \Omega$

$X_{\mathrm{m}} = 176\ \Omega$

Calculate the input current, torque and efficiency of the machine when it is operating with a slip of 0.044. Windage and friction losses may be neglected.

From parameter values given

$$\frac{R_r'}{2s} = \frac{9.6}{0.088} = 109.09\ \Omega$$

$$\frac{R_r'}{2(2-s)} = \frac{9.6}{3.912} = 2.45\ \Omega$$

$$\frac{X_r'}{2} = 5.9\ \Omega$$

$$\frac{X_m}{2} = 88\ \Omega$$

Hence

$$\boldsymbol{Z}_f = \frac{(109.09 + 5.9j) \times 88j}{109.09 + 93.9j} = 66.81\angle 52.39^\circ = 40.31 + 53.28j\ \Omega$$

and

$$\boldsymbol{Z}_b = \frac{(2.454 + 5.9j) \times 88j}{2.454 + 93.9j} = 5.99\angle 68.92^\circ = 2.15 + 5.59j\ \Omega$$

Total impedance seen from the machine terminals is then:

$$\boldsymbol{Z} = \boldsymbol{Z}_f + \boldsymbol{Z}_b + 8 + 11.4j = 50.46 + 70.27j = 86.51\angle 54.32^\circ\ \Omega$$

With $\boldsymbol{V}$ as reference the rotor current is then:

$$\boldsymbol{I}_r' = \frac{\boldsymbol{V}}{\boldsymbol{Z}} = 2.77\angle -54.32^\circ = 1.62 - 2.25j\ \text{A}$$

The input current is then

$$\boldsymbol{I}_s = \boldsymbol{I}_r' + \frac{240}{2400} = 1.72 - 2.25j = 2.83\angle -52.62^\circ\ \text{A}$$

Now, for the forward rotor circuit:

$$\boldsymbol{V}_f = \boldsymbol{I}_r' . \boldsymbol{Z}_f = 185.1\angle -1.93^\circ\ \text{V}$$

and

$$\boldsymbol{I}_f' = \frac{\boldsymbol{V}_f}{109.25\angle 3.1^\circ} = 1.69\angle -5.03^\circ\ \text{A}$$

when

$$P_f = 1.69^2 \times 109.09 = 311.6\ \text{W}$$

For the backward rotor circuit:

$$\boldsymbol{V}_b = \boldsymbol{I}_r' . \boldsymbol{Z}_b = 16.59\angle 14.6^\circ\ \text{V}$$

and

$$\boldsymbol{I}_b' = \frac{\boldsymbol{V}_b}{6.39\angle 67.42^\circ} = 2.6\angle -52.82^\circ\ \text{A}$$

then

$$P_b = 2.6^2 \times 2.454 = 16.59\ \text{W}$$

Hence, power out

$$P_{out} = (311.6 - 16.59) \times (1 - s) = 282 \text{ W}$$

when

$$\text{Torque} = P_{out} \cdot \frac{2}{(1-s)\omega} = 1.88 \text{ N m}$$

Input power is given by:

$$P_{in} = 311.6 + 16.59 + \frac{240^2}{2400} + 2.77^2 \times 8 = 413.6 \text{ W}$$

Therefore the motor efficiency is:

$$\eta = 100 \times \frac{282}{413.6} = 68.2\%$$

6.4.1 *Starting single-phase induction motors*

In order to start a single-phase induction motor the auxiliary winding is used to increase the magnitude of the forward field with respect to the backward field. The auxiliary winding is usually wound on the stator at an angle of $\frac{\pi}{2}$ electrical radians relative to the main winding.

The phase shift between the currents in the main and auxiliary windings is typically produced either by means of adding resistance to the auxiliary winding as in Fig. 6.32(a) – **split-phase starting** – or by means of a capacitor as in Fig. 6.32(b). The auxiliary winding is usually short time rated and switched out by means of a centrifugal switch mounted on the rotor shaft at around 70% of full load speed.

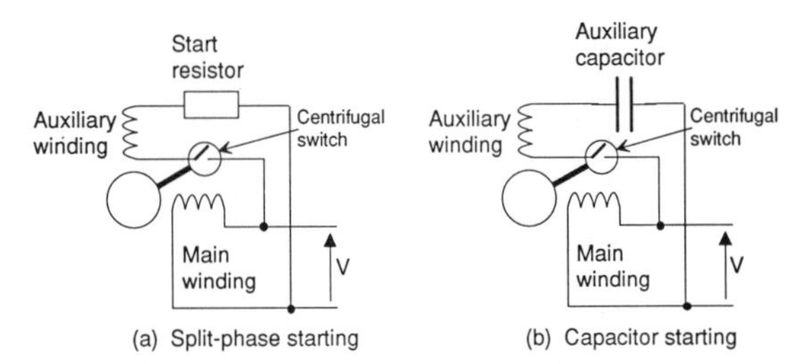

Fig. 6.32 Starting techniques for a single-phase induction motor: (a) split-phase starting; and (b) capacitor starting.

Comment

For a capcitor-start/capacitor-run single-phase motor the run capacitor and the auxiliary winding will be permanently connected and the centrifugal switch is therefore omitted.

Typical ratings for single-phase induction machines are less than 10 kW for capacitor-start and capacitor-start/capacitor-run motors and around 1 kW for split-phase start motors.

Single phase induction motors are widely used for domestic applications such as refrigerator compressors, washing machine pumps, central heating pumps and fans.

Exercises

1 A three-phase, 60 Hz, six-pole 440 volt (line), star connected wound rotor induction machine has the following per phase parameters:

Stator resistance = Negligible
Stator leakage inductance = 0.003 H
Rotor resistance = 0.3 Ω
Rotor leakage inductance = 0.002 H
Magnetizing inductance = 0.1 H
Core loss resistance = Infinite
Turns ratio (N_s/N_r) = 1.2:1

(a) What is the maximum torque of the machine?
(b) What would be the torque developed at a slip of 0.08 with a shorted rotor?
(c) What is the starting torque with a shorted rotor?
(d) The machine is required to drive a load of torque of 100 N m at 720 rev/min. What value of resistance must be added to each phase of the rotor?
(e) What value of resistance must be added to each phase of the rotor to give maximum starting torque?
(f) At what speed would the machine drive a load of torque 100 N m with the added resistance of (e)?
(g) If the supply frequency is now changed to 50 Hz with the same value of supply voltage, what will be the new maximum torque for the machine?
(h) What torque would be developed at a slip of 0.08 with the supply conditions of (g)?
(i) What power would be fed back to the 60 Hz supply if the machine is driven at a slip of −0.04 with no added rotor resistance?

A three-phase, 50 Hz, six-pole, 415 volt star connected squirrel cage induction motor has a stator resistance of 0.3 Ω per phase. The no-load and locked rotor tests on the machine gave the results in Table 6.1.

Table 6.1 Results of no-load and locked rotor tests

Test	V (line) (V)	I (A)	Power (W)
No-load	415	8	1200
Locked rotor	190	40	6000

Using these test results estimate the equivalent circuit values for the machine and hence calculate the starting torque of the machine.

3 A three-phase, four-pole, 50 Hz, 380 volt (line) induction motor develops 5 kW when running at 1420 rev/min. Windage and friction losses are estimated at 240 W. Calculate the rotor copper loss and the airgap power for the machine.

4 A three-phase, 50 Hz, 415 volt (line) induction motor rated at 18 kW (mechanical) with a full load efficiency of 87% is to be started by means of a star–delta starter. With full load voltage applied with the stator connected in delta the locked rotor line current in the stator is 165 A. Neglecting the effect of magnetizing current determine the ratio of full load current to starting current for the machine.

5 A three-phase, 50 Hz, 1 kV (line), six-pole induction machine has the following per phase parameters:

Stator resistance	$=0.4\ \Omega$	Rotor resistance	$=1.1\ \Omega$
Stator leakage inductance	$=0.0028$ H	Rotor leakage inductance	$=0.002$ H
Core loss resistance	= Infinite	Turns ratio (N_s/N_r)	$=1.1:1$
Magnetizing inductance	$=0.12$ H	Mechanical losses	$=1500$ W

and is used as an induction generator in association with a windmill. The windmill is rated at 50 kW (electrical) at a windspeed of 16 km/h when the windmill blades are rotating at 15 rev/min. (a) Given that the windmill extracts energy from the wind according to the relationship

$$\text{Power developed} \propto (\text{windspeed})^3$$

find the slip and shaft speed of the generator when it is supplying 50 kW to the 1 kV (line), 50 Hz busbar at a windspeed of 16 km/h (neglect stator losses). (b) Hence find the gearbox ratio required between the windmill and the shaft of the machine (c) Assuming that the gearbox has an efficiency of 96% calculate the lowest operating windspeed for the windmill.

6 A 240 volt, 50 Hz, four-pole single phase induction motor has the following parameters:

R_1	$=1.2\ \Omega$	X_m	$=56\ \Omega$
R_2	$=3.8\ \Omega$	Turns ratio (N_s/N_r)	$=1:1$
X_1	$=2.8\ \Omega$	Core loss	$=40$ W
X_2	$=3\ \Omega$	Windage and friction	$=15$ W

If the motor is operated at a slip of 0.06, what will be the output power and torque?

7 A four-pole, 50 Hz, 120 volt single phase induction motor has the following parameters:

Stator resistance	$=1.7$ W	Rotor resistance	$=3.6$ W
Stator leakage reactance	$=2.4$ W	Rotor leakage reactance	$=2.4$ W
Magnetizing reactance	$=54$ W	Turns ratio (N_s/N_r)	$=1:1$

Determine the ratio of forward voltage to backward voltage when the motor is operating at a slip of 0.05. Using this relationship plot the positions of the forward and backward travelling vectors in the airgap for the period 0 to 20 ms and hence find the locus of the resultant airgap flux vector.

Other electrical machine types 7

7.1 Single-phase a.c. commutator motor

The single-phase a.c. commutator motor is connected as shown in Fig. 7.1 and is essentially a series d.c. motor running on an a.c. supply. As the field and armature current are reversed simultaneously by the a.c. current the torque direction remains constant throughout. The magnitude of the torque will, however vary from zero to a maximum in the course of each half-cycle of armature current.

In order to operate effectively from the a.c. supply, the design of the a.c. commutator motor differs from that of the d.c. motor to take account of the reactance effects present when connected to an a.c. supply. In particular, for machines of similar rating the number of turns on the stator winding of the a.c. commutator motor is reduced, reducing the stator inductance. The number of turns on the armature is then increased to compensate for the reduction in the stator field.

The commutation performance of the a.c. commutator motor is also worse than that of a d.c. series motor and compensation for armature reaction effects may well be required to improve performance. This typically takes the form of a compensating winding on the stator connected in series with the main field winding and with its axis at right angles to that of the main field winding. The presence of the compensating winding not only reduces the armature reaction but also results in an improvement in operating power factor. Alternatively, a short-circuited winding can be introduced on to the stator with its axis between that of the field and armature windings. Currents are induced into this winding by linkage with the main field flux generate a compensating m.m.f. along the axis of the armature winding.

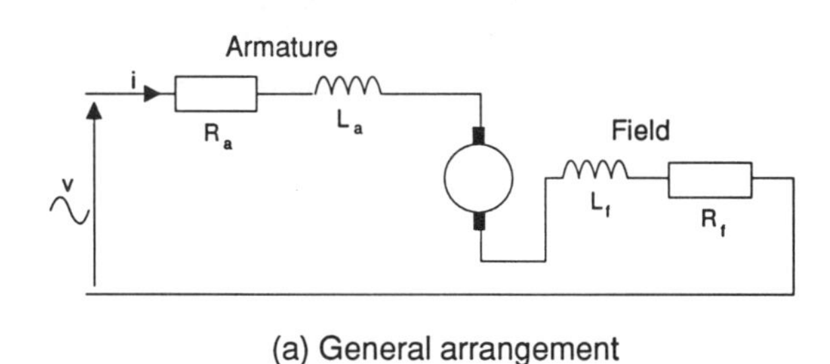

(a) General arrangement

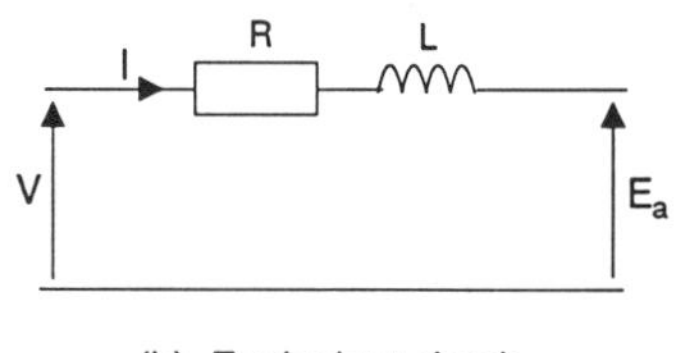

(b) Equivalent circuit

Fig. 7.1 Single-phase commutator motor: (a) general arrangement; and (b) equivalent circuit.

Comment

A series commutator motor capable of running on both a.c. and d.c. is referred to as a **series universal** motor.

For the machine of Fig. 7.1:

$$e_a = K_{scm} \cdot i_a \cdot \omega_m \tag{7.1}$$

$$v = (R_a + R_f) \cdot i_a + (L_a + L_f) \cdot \frac{di_a}{dt} + e_a \tag{7.2}$$

and

$$T = K_{scm} \cdot i_a^2 \tag{7.3}$$

Neglecting the speed variations resulting from the pulsating torque in the machine these equations can be rewritten as phasor equations using r.m.s. values as:

$$\boldsymbol{E}_a = K_{scm} \cdot \boldsymbol{I}_a \cdot \omega_m \tag{7.4}$$

$$\boldsymbol{V} = [(R_a + R_f) + j\omega(L_a + L_f)]\boldsymbol{I}_a + \boldsymbol{E}_a = (R + jX)\boldsymbol{I}_a + \boldsymbol{E}_a$$
$$= \boldsymbol{I}_a[(R + K_{scm} \cdot \omega_m) + jX] \tag{7.5}$$

and

$$T = K_{scm} \cdot \boldsymbol{I}_a^2 \tag{7.6}$$

7.1.1 *Speed control of a.c. commutator motors*

The simplest means of achieving speed control of an a.c. commutator motor is that shown in Fig. 7.2 in which a single thyristor is used to provide half-wave rectification with speed controlled by varying the firing angle of the thyristor. The presence of the freewheeling diode reduces torque fluctuations by maintaining a continuous current in the machine. Speed stability is also improved by increasing the firing delay angle (α

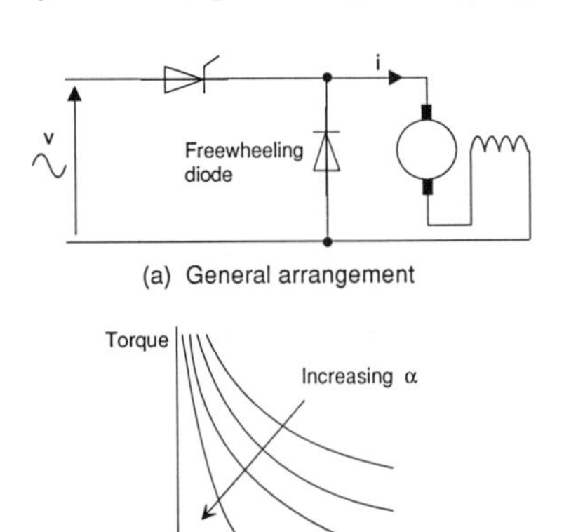

Fig. 7.2 Speed control of an a.c. series commutator motor using a single thyristor: (a) general arrangement; and (b) generalized torque/speed characteristic.

necessary to achieve a given reduction in speed which means that the performance is less affected by any small variations in the firing angle.

By using a full-wave controller based on a thyristor bridge or triac as in Fig. 7.3 the d.c. component in the current drawn from the a.c. supply that is present with the arrangement of Fig. 7.2 is eliminated. Improvements in machine performance also occur as a result of the reduction in the current variation in the machine windings during operation.

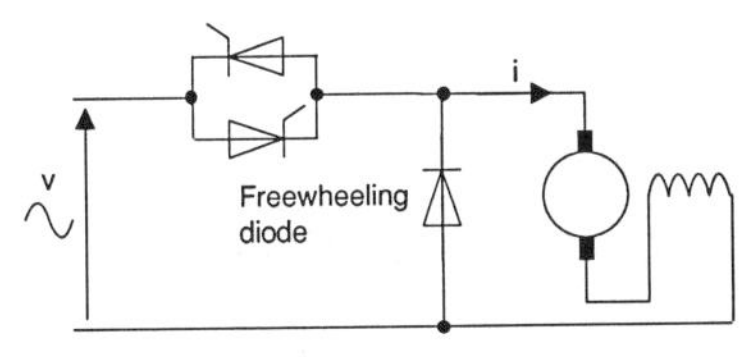

Fig. 7.3 Speed control of an a.c. series commutator motor using a back-to-back thyristor pair.

Example 7.1 a.c. commutator motors
A 220 volt, 50 Hz universal motor is rated at 0.25 kW has the following parameters:

Combined resistance of field and armature $= 11.6\ \Omega$
Armature inductance $= 0.44$ H
Machine constant K_{scm} $= 0.502$

If the mechanical losses when running on no load at 9200 rev/min are 110 W and are proportional to (speed)^2 find the torque developed, mechanical power output, supply current and efficiency when the motor is operating at 4800 rev/min from a 220 volt, 50 Hz supply.

$$E_a = K_{scm} . I_a . \omega_m = 0.502 . I_a . \omega_m$$

and

$$V = I_a(R + jX) + E_a = I\left(11.6 + 0.502 \frac{2 . \pi . 4800}{60} + 138.2j\right)$$

Hence

$$I_a = 0.738\underline{|-27.64^\circ} \text{ A}$$

$$T_e = K_{scm} . I_a^2 = 0.502 \times 0.738^2 = 0.274 \text{ N m}$$

$$\text{Equivalent power} = 0.274\left(\frac{2 . \pi . 4800}{60}\right) = 137.6 \text{ W}$$

$$\text{Mechanical losses} = 110\left(\frac{4800}{9200}\right)^2 = 29.9 \text{ W}$$

$$\text{Power out} = 137.6 - 29.9 = 107.7 \text{ W}$$

$$\text{Losses} = 29.9 + 0.738^2 \times 11.6 = 36.2 \text{ W}$$

$$\text{Efficiency} = \frac{100 \times \text{Power out}}{\text{Power out} + \text{Losses}}\% = \frac{100 \times 107.7}{107.7 + 36.2}\% = 74.8\%$$

7.2 Stepper motors

Stepper motors can be used to provide either a continuous controlled rotation or a series of discrete angular motions, making them very suitable for use in applications involving computer control of motion. A variety of different types of stepper motor are available offering a range of characteristics.

7.2.1 *Variable reluctance stepper motor*

The construction of a simple, single stack, 12-step, variable reluctance stepper motor is shown in Fig. 7.4. In this form, the stator carries three separate windings (phases) on six equally spaced poles while the rotor has only four poles and carries no windings.

With phase AA′ energized, the rotor will be pulled into the position shown in Fig. 7.4(a), with one pair of poles aligned with the phase A and A′ poles on the stator. If phase AA′ is now turned off and phase BB′ energized, the rotor will move to the position shown in Fig. 7.4(b). A further rotation to the position of Fig. 7.4(c) occurs when phase BB′ is turned off and phase CC′ energized. Repeating the sequence of phases AA′–BB′–CC′–AA′, etc., will cause the rotation to continue with a total of 12 steps required before the rotor returns to its original position. If this sequence in which the stator phases are energized is reversed to give AA′–CC′–BB′–AA′, etc., the direction of rotation of the rotor will be reversed.

A feature of this form of construction is the ability to generate intermediate steps by energizing pairs of phases together. Thus, for the motor shown, energizing phases AA′ and BB′ together will cause the rotor to assume a position halfway between those of Figs 7.4(a) and 7.4(b).

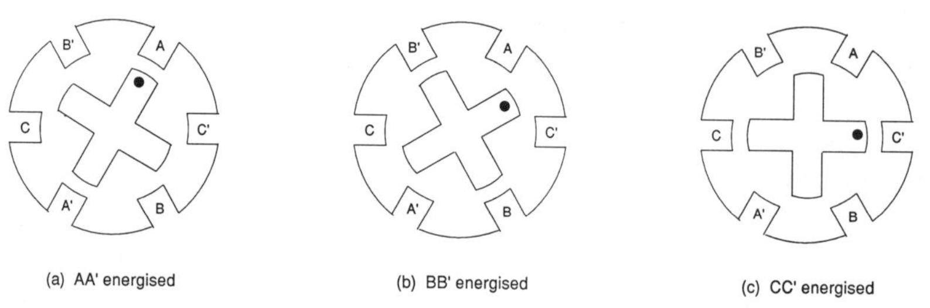

Fig. 7.4 Single stack variable reluctance stepper motor: (a) AA′ energized; (b) BB′ energized; and (c) CC′ energized.

Comment

This approach may be used to achieve larger numbers of steps in a single revolution than are provided by the basic motor.

An alternative form of construction for a variable reluctance stepper motor is the multi-stack form of Fig. 7.5. Here, each of the stator phases is placed on a different

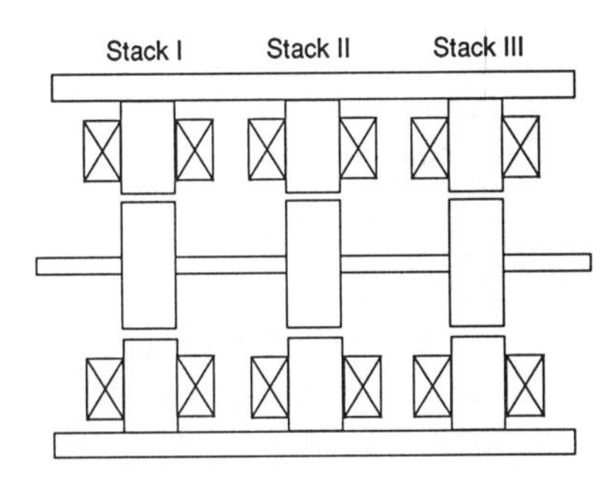

(a) Three stack variable reluctance motor

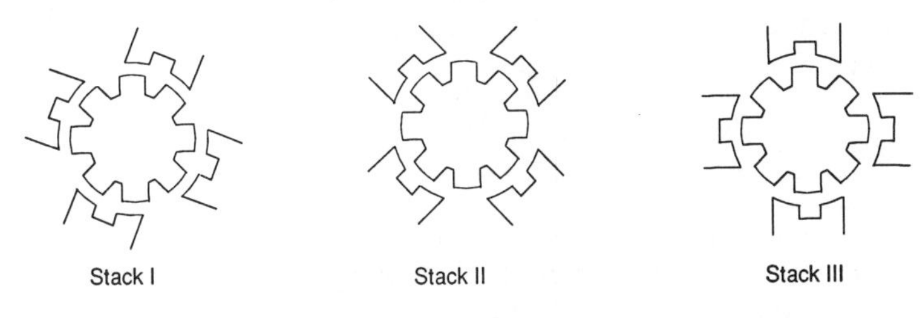

(b) Rotor stacks, stack I energised

Fig. 7.5 (a) Four-pole, three-stack variable reluctance stepper motor. (b) Rotor stacks (stack I energized): (i) Stack I; (ii) Stack II; and (iii) Stack III.

stator stack and each stack is associated with its own rotor stack as shown. Each of these rotor stacks is aligned at some angle with respect to the other rotor stacks, enabling a smaller step angle to be achieved.

The relationship between the number of stator poles, the number of rotor poles, the number of phases and the step angle for the variable reluctance stepper motor is given by:

$$\text{Step angle} = \theta_s = \frac{360}{s \cdot n} \tag{7.7}$$

where s is the number of stacks (phases) and n is the number of teeth per stack (phase) = number of poles × teeth per pole.

Comment

For the machine of Fig. 7.5:

$$s = 3$$

and

$$n = 4 \times 2 = 8$$

Hence

$$\theta_s = \frac{360}{24} = 15°$$

It should be noted that a variable reluctance stepper motor will only produce a torque when the stator windings are energized and that the rotor will otherwise be free to rotate if a torque is applied to the shaft.

7.2.2 *Permanent magnet stepper motors*

The permanent magnet stepper motor of Fig. 7.6 has, as its name suggests, a permanent magnet embedded in its rotor which increases the flux in the machine and also results in the provision of a holding or detente torque on the rotor when the stator windings are de-energized.

As with the variable reluctance stepper motor, energizing the stator windings in turn results in a stepped rotation of the rotor. The step angle is determined by the rotor and stator teeth according to:

$$\text{Step angle} = \theta_s = \frac{360}{m\,.\,n} \tag{7.8}$$

where m is the number of phases and n is the number of teeth. The maximum number of steps per revolution that can be achieved is limited by the construction of the machine and the size of the stator and rotor teeth that can be used.

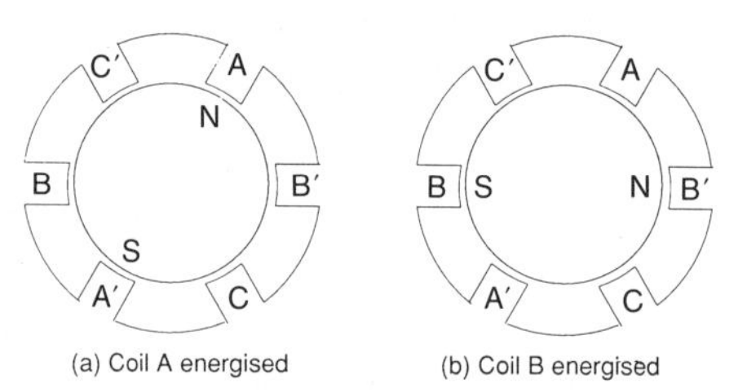

Fig. 7.6 Permanent magnet stepper motor: (a) coil A energized; and (b) coil B energized.

7.2.3 *Hybrid stepper motors*

The hybrid stepper motor combines features of both the variable reluctance and the permanent magnet stepper motors in that it combines a permanent magnet core and a toothed rotor. Hybrid stepper motors generally operate with smaller step angles than the variable reluctance and permanent magnet stepper motors and they also offer a higher torque-to-volume ratio than these. A hybrid stepper motor will also provide a détente torque when its stator windings are de-energized.

7.2.4 *Stepper motor operation*

The generalized torque/stepping-rate characteristic for a stepper motor is shown in Fig. 7.7 and exhibits a number of different regions and boundaries.

(a) *Pull-in curve*

The pull-in curve defines those combinations of torque and stepping rate against which a motor can start or stop without losing steps. The shape of the pull-in curve is influenced by the inertia of the load that is to be started and increasing the load inertia will reduce the stepping rate to be used if steps are not to be missed during acceleration.

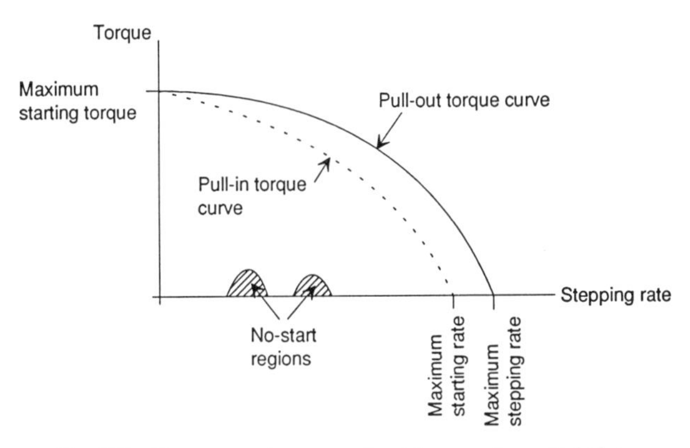

Fig. 7.7 Stepper motor generalized torque characteristics.

(b) *Pull-out curve*

The pull-out curve defines the torque/stepping rate relationship for the motor at steady speed and sets the limiting conditions for operation if steps are not to be lost. The shape of the pull-out curve is influenced by the nature and type of drive circuit used by the motor. It is also possible that the shape of the pull-out curve will be distorted in the manner of Fig. 7.8 as a result of internal resonance conditions in the motor arising from its construction.

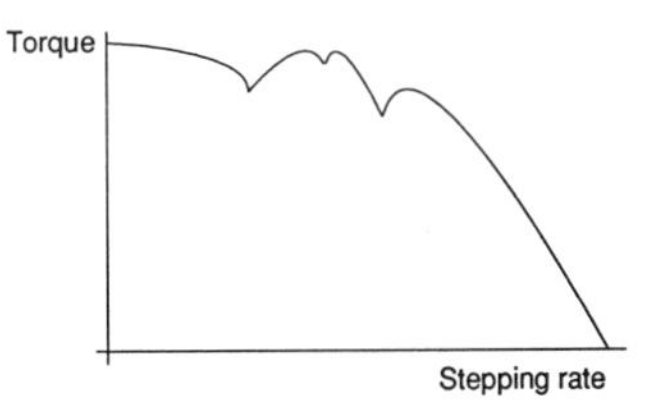

Fig. 7.8 Distorted stepping motor pull-out torque characteristic.

(c) *No-start regions*

No-start regions are, as their name suggests, regions in the torque/stepping rate characteristic defining conditions against which the motor is unable to start. These regions are set by the construction of the motor and the type of drive used.

(d) *Inertia affects*

The operation of a stepper motor is significantly influenced by the inertia of the driven load. In particular, on starting the load inertia influences the shape of the pull-in characteristic while on stopping, the rate of deceleration must be controlled in relation to the load inertia in order to prevent any overshoots occurring. Appendix B deals with the calculation of the inertia of different types of load.

7.2.5 *Stepper motor drive circuits*

A typical drive circuit for one phase of a stepper motor is shown in Fig. 7.9. The base signals of the transistor are derived from the control circuit, which may be a dedicated integrated circuit or a microprocessor, via a suitable interface and are used to switch the transistor into the on state.

As the performance of the stepper motor is improved by having the stator current reach its full value as rapidly as possible the time constant (τ) of the stator circuit is reduced and the rate of rise of current is increased by introducing the **forcing resistor** (R_1) in series with the stator since:

$$\tau = \frac{L}{R} \tag{7.9}$$

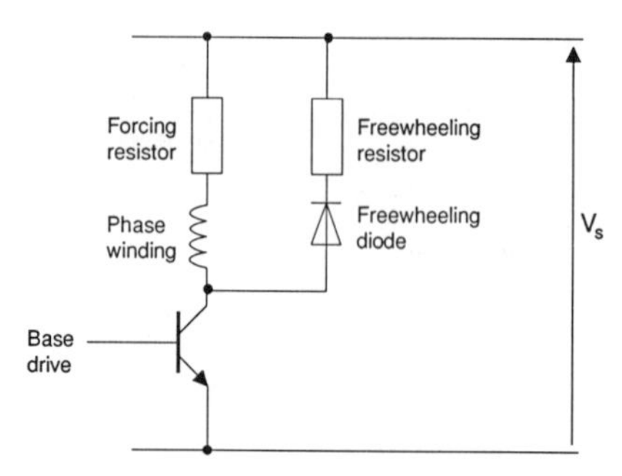

Fig. 7.9 Drive circuit for one phase of a variable reluctance stepper motor.

In order to ensure that the correct final value of stator current is obtained this in turn requires an increase in the applied voltage such that:

$$I_{phase} = \frac{V_s}{R_{phase} + R_1} \tag{7.10}$$

On turn-off of the transistor, the energy stored in the magnetic field of the stator winding is dissipated through the forcing resistor and the **freewheeling resistor** (R_2) via the diode, causing the rapid collapse of the stator current.

Where a bidirectional current drive is required the bridge circuit of Fig. 7.10 can be used with the transitors switched in pairs (T_1 and T_2, and T_3 and T_4).

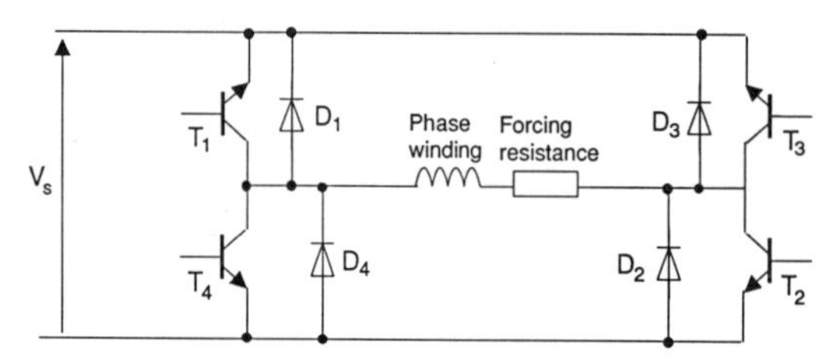

Fig. 7.10 Bi-directional drive circuit.

Example 7.2 Stepper motor operation
A variable reluctance stepper motor has the following per phase parameters:

Inductance $= 12$ mH
Resistance $= 32\ \Omega$
Rated current $= 1.2$ A

If 99% of full rated current is to be achieved within 1 ms of turn-on find the value of the forcing resistance and source voltage required. On turn-off, the current is required to collapse within 0.8 ms, hence find the value of the freewheeling resistor required.

Assuming an exponential raise of current then 99% of full rated curent will be reached after a period of 5 times the system time constant.

$$\text{Hence effective time constant on turn-on} = \frac{0.001}{5} = \frac{L}{R_{tot;1}}$$

When

$$R_{tot;1} = 60\ \Omega$$

Then

$$R_1 = 60 - 32 = 28\ \Omega$$

The required voltage is then:

$$V_s = 1.2 \times 60 = 72\ \text{V}$$

$$\text{Effective time constant on turn-off} = \frac{0.0008}{5} = \frac{L}{R_{tot;2}}$$

Hence

$$R_{tot;2} = 75\ \Omega$$

when

$$R_2 = 75 - 60 = 15\ \Omega$$

7.3 Switched reluctance motor

The switched reluctance motor has essentially the same configuration as a single stack variable reluctance stepper motor but is designed for continuous rotation at high powers and torques.

Referring to Fig. 7.11, continuous operation requires the application of current to each of the stator phases in turn at a rate which is dependant on and determined by the variation of rotor position with time. The timing of the firing of the controlling switching devices is determined by reference to the rotor position by means of either Hall effect or optical sensors. These provide input data for the control algorithms which then compute the required firing angles.

Switched reluctance motors are high efficiency machines capable of full four-quadrant operation. They are of simple construction with no rotor windings and have a flexible operating characteristic. Indeed, by varying the phase relationship of the applied current with respect to the rotor position the characteristic of the switched reluctance motor can be shaped to a significant degree enabling it, for instance, to assume the shaft characteristics of either a d.c. series or a d.c. shunt motor, resulting in their being used for traction applications. They do however require a relatively complex controller.

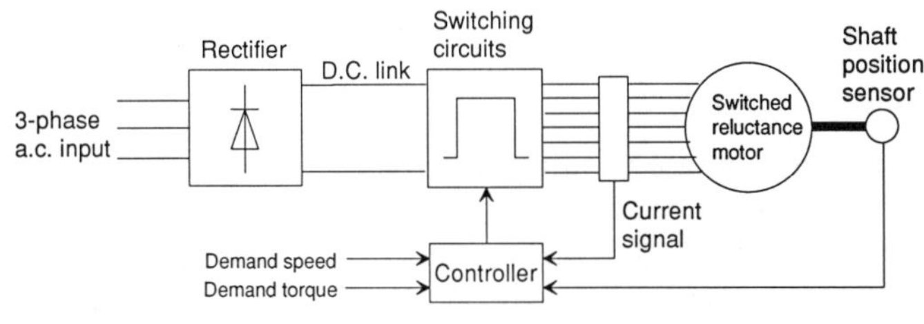

Fig. 7.11 Switched reluctance motor controller schematic.

7.3.1 *Speed control of switched reluctance motors*

The control strategy adopted for a switched reluctance motor depends upon the operating regime. Thus for starting and low speeds, control would be applied to limit the stator current. As the machine accelerates, the stator current will fall as a back e.m.f. appears in the windings and firing angles are adjusted to maintain the current between pre-set limits during run-up.

In the mid-range the controller provides for control of both thyristor turn-on and turn-off before finally establishing operation with controlled turn-on only at high speeds.

7.4 Disc motor

A disc motor is a homopolar d.c. machine in which the rotor consists of a thin disc of non-magnetic, non-conducting material. The rotor conductors are then arranged radially on each side of the disc and may be formed either as discrete windings or printed on to the rotor disc. Current is transferred to the rotor by axially mounted brushes close to the shaft of the machine.

Comment

The form of disc motor with the rotor conductors printed on to the rotor disc is referred to as a **printed circuit motor**.

Operation of the disc motor is exactly as for a conventional d.c. machine. However, special features associated with the form of construction are:

- very low inertia resulting in a high torque/inertia ratio;
- negligible armature reaction giving a near linear speed/torque characteristic;
- a large number of armature conductors resulting in little torque ripple even at low speeds;
- a high overload current withstand capability.

The low inertia of the disc motor has resulted in their extensive use as servomotors, particularly with the incorporation of a built in encoder of some form. Other uses have included a variety of drive applications such as lawn-mowers and electric vehicles in which the motor was built into the wheel hub.

7.5 Linear motors

7.5.1 *Linear induction motor*

If a conventional three-phase winding is flattened out, the result will be a linearly travelling magnetic field as opposed to the more usual rotating magnetic field. If this winding is now placed alongside a plate of conducting material the travelling field will interact with the eddy currents induced in the plate to produce a force on the plate in the direction of the travelling field. This is the basis of the linear induction motor.

Comment

By fixing the plate, which then becomes the reaction member, the effect is to transfer the motion to the stator. This is the normal mode of operation.

Typical constructions are the short stator and short secondary forms shown in Figs 7.12(a) and 7.12(b). The double sided short stator of Fig. 7.12(c) eliminates the need for the presence of magnetic material in the secondary and also tends to balance out the side forces present in a single sided system. Applications include traction schemes and Fig. 7.13 shows a sectional view of the installation on the Maglev passenger transit system at Birmingham International Airport in which the secondary or reaction rail consists of an aluminium plate mounted on a steel beam and produces a tractive force of 2 kN for a steady speed of 15 m/s and up to 4 kN during acceleration or deceleration.

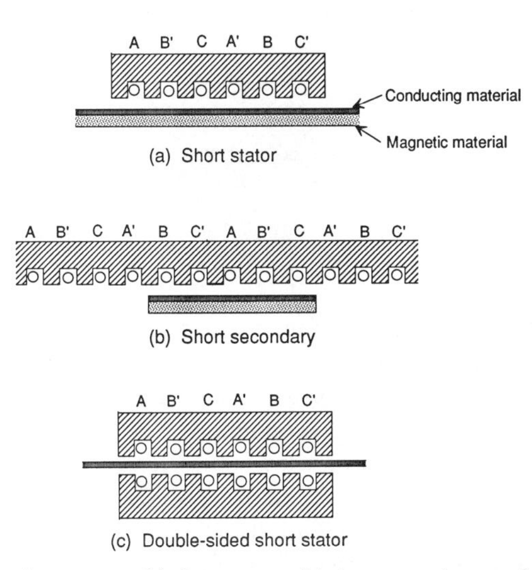

Fig. 7.12 Linear induction motors: (a) short stator; (b) short secondary; and (c) double-sided short stator.

7.5.2 *Linear d.c. motor*

If a current is passed through the reaction member by means of brushes mounted on the rail then this current can be made to react with a d.c. field produced by the stator to produce linear motion.

Comment

The source of the field in a linear motor is referred to as the stator even though it might form the moving element when operated with a fixed reaction rail or secondary as in the case of the Maglev system referred to above.

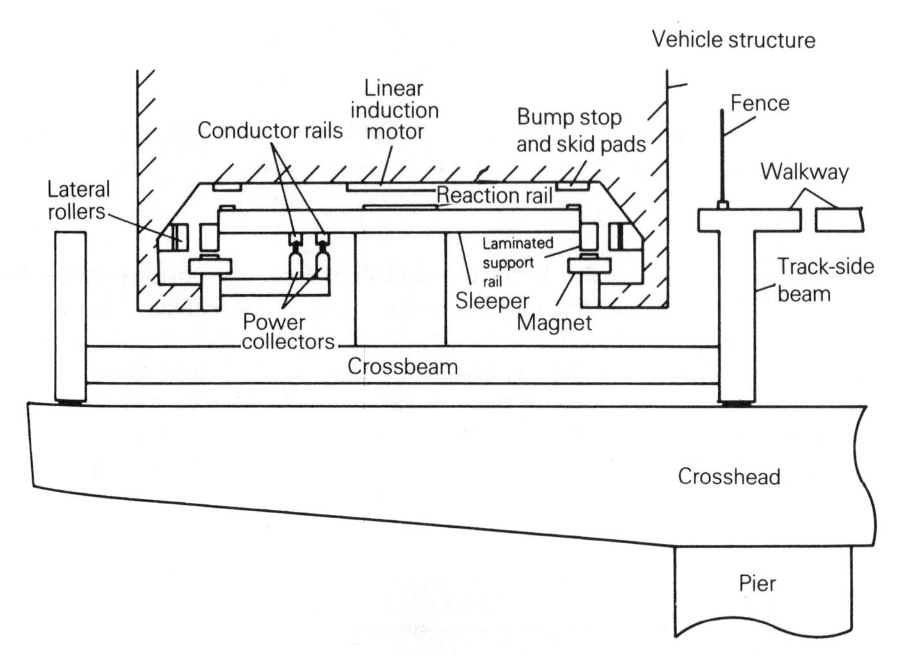

Fig. 7.13 Maglev actuator. (Reprinted courtesy of Van Nostrand Rheinhold Ltd.)

7.6 Micromotors

The development of the techniques for the micromachining of silicon have enabled the production of a range of capacitive stepper motors at the micrometre scale. Construction is similar to that of the single stack variable reluctance motor of Fig. 7.4 but with a much higher number of poles on both stator and rotor. Energization of pairs of stator poles moves the adjacent rotor poles into alignment as a result of the capacitive coupling between the stator and the rotors. A typical rotor diameter for such a machine would be of the order 500 μm to 600 μm.

A variation on the simple capacitive micromotor is the **nutating** or **wobble motor** of Fig. 7.14. Here, the rigid outer ring is attracted to the stator electrodes causing it to wobble vertically about its centre. Connection of this outer ring to a constrained inner ring by means of flexible elements results in a pure rotational motion of this inner ring about its centre. As the outer ring is effectively rolling along a path shorter than its own circumference, the speed of rotation of the inner ring is slow in relation to the rate of switching between stator segments. The reduction ratio between the contact point and

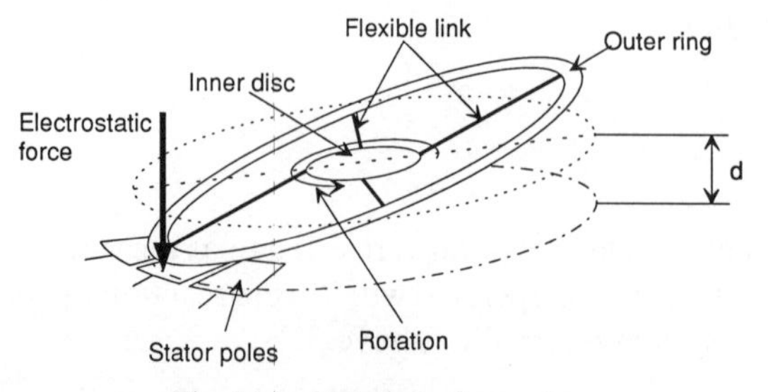

Fig. 7.14 Nutating micromotor.

the inner rotor for a rotor of radius R_0 is then:

$$\text{Ratio} = \frac{R_0 - R_{\text{path}}}{R_{\text{path}}} \tag{7.11}$$

where

$$R_{\text{path}} = (R_0^2 - d^2)^{\frac{1}{2}}$$

and d is the vertical separation between the centre of the rotor and the stator.

Using this technique reduction ratios of the order of 20 000:1 have been achieved with a ring diameter of around 200 μm to 300 μm.

7.7 Ultrasonic motors

Ultrasonic motors use a piezoelectric resonator operating at ultrasonic frequencies to drive a passive rotor by means of a friction contact. Such motors are distinguished by the absence of a magnetic field, high torques at low speeds and a large holding torque when de-energized.

Comment

When an electric field is applied to a piezoelectric material, the material will undergo a dimensional change which is related to the magnitude and orientation of the applied field. This is the **inverse piezoelectric effect**.

In comparison with electromagnetic motors ultrasonic motors offer higher powers per unit volume and can generate high torques at low speeds without the need for a gearbox. Additionally, the mechanical inertia of the rotor is low giving a fast response which is capable of being precisely controlled by the drive electronics.

The main disadvantage of the ultrasonic motor, and the primary limitation on an operational life of the order of a few hundred hours, is the mechanical wear that occurs between the resonator and the rotor. The introduction of improved surface treatments to reduce the wear rate will be a major development in extending the life of such machines.

Applications for ultrasonic motors centre around precise positioning applications requiring intermittent operation and include autofocus lens drives such as that shown in Fig. 7.15, printers and plotters.

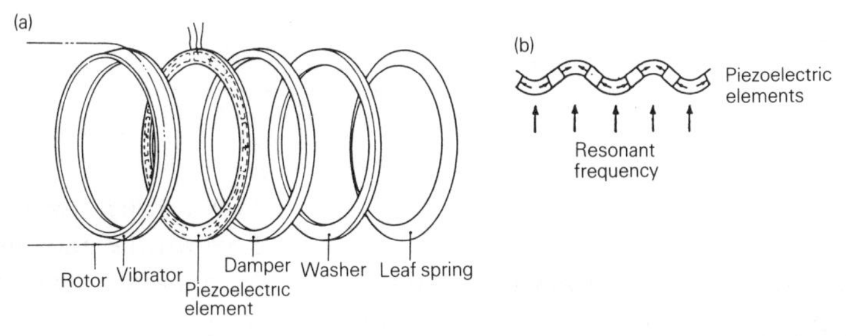

Fig. 7.15 Ultrasonic motor: (a) exploded diagram; (b) bending wave produced by the piezoelectric elements.

8 Mechanical drives

While electrical machines provide the great bulk of industrial drives, in many applications a mechanical alternative may be a better choice. For instance, where there is a particular risk of fire or explosion a pneumatic drive may well be preferred while hydraulic drives offer a significantly greater power-to-weight ratio, particularly when providing high torques at low speeds. In this context, Fig. 8.1 illustrates in a simplified form the overlaps in the operating regions for various types of drive.

In this chapter some of these alternative forms of drive are briefly introduced for comparison with the electrical drives considered in the earlier chapters.

8.1 Pneumatic motors

The two basic types of pneumatic motor in common use are the vane motor and the piston motor, examples of which are shown in Figs 8.2 and 8.3, respectively. The main advantages of pneumatic motors are as follows.

- They are intrinsically safe as they use air as their power source.
- There is an inherent torque limit set by the available supply pressure.
- They are robust and not particularly vulnerable to mechanical shock or jamming, especially when directly connected.

The main areas of disadvantage are as follows.

- They are very inefficient in terms of energy utilization with as little as 20% of the input energy available as output at the shaft.
- They are noisy and generally require an exhaust silencer.
- Speed is highly load dependent and control of speed can only be achieved by throttlng the air supply to the motor.

8.2 Hydraulic motors

Hydraulic motors offer a high torque-to-volume and power-to-weight ratios and are therefore used where torques are to be transmitted within a restricted space. A variety of different types of motor are available covering a range of applications as in Table 8.1.

Table 8.1 Applications of hydraulic motors

Motor type	Application
Gear; vane	Low torque, low power
Vane; axial piston	Low torque, medium power
Axial piston; radial piston (inward working)	Medium torque and/or power
Axial piston; radial piston (outward working)	Very high torque, high power

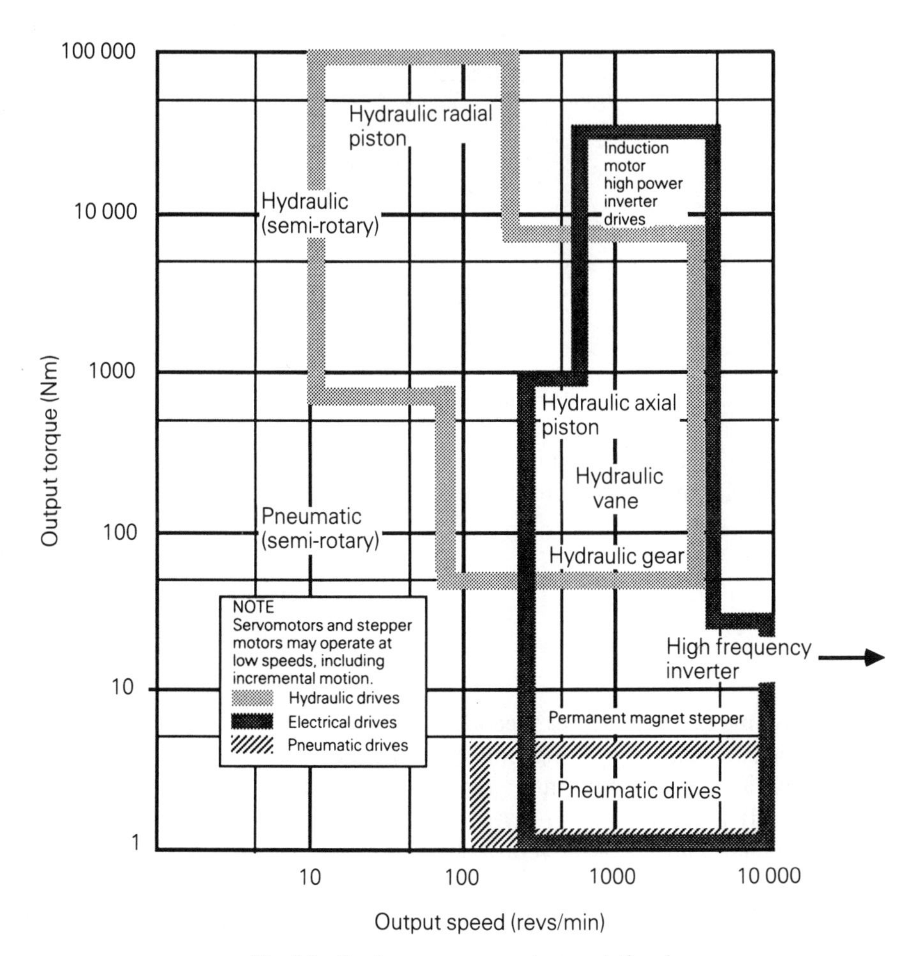

Fig. 8.1 Service energy converters – rotational.

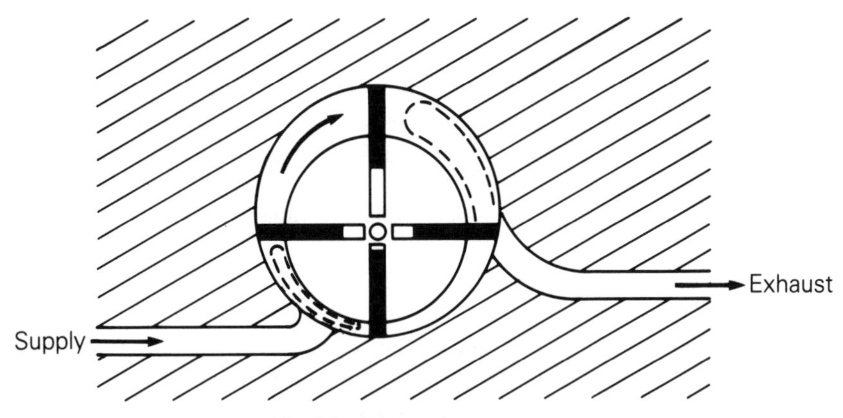

Fig. 8.2 Pneumatic vane motor.

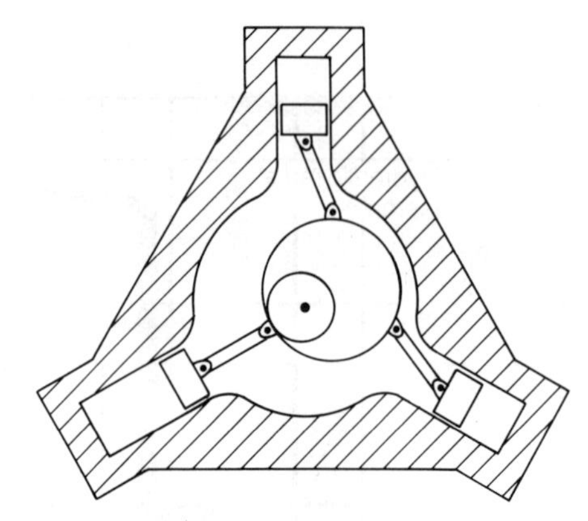

Fig. 8.3 Pneumatic piston motor.

The torque and speed relationships for a hydraulic motor are expressed in terms of their specific displacement according to the relationships:

$$T = K_h . D . p \tag{8.1}$$

and

$$Q = K_h . D . \omega \tag{8.2}$$

where:

Q is the volume flow rate (litres^3/second);
K_h is a constant defined by the geometry of the motor;
D is the specific displacement in m^3/radian of the motor;
p is the supply pressure;
ω is the angular velocity of the motor shaft.

Comment

Equations (8.1) and (8.2) should be compared with equations (4.13) and (4.8) for a d.c. machine.

All hydraulic motors depend for their operation on the close working clearances produced during manufacture and the maintenance of these clearances during their operational lifetime. As all hydraulic motors are subject to internal leakage, their internal displacement is always less than the ideal and the effect of this internal leakage is therefore to reduce the output speed below that which would be expected from the geometric displacement. This is expressed as a hydraulic efficiency in the form:

$$\eta_H = \frac{\omega_{actual}}{\omega_{ideal}} \tag{8.3}$$

According to the type of motor the value of η_H will vary from around 75% for small gear motors to over 95% for axial piston machines and tends to reduce with an increasing pressure differential across the motor and is also lower at lower speeds. Mechanical losses vary according to the type of motor but these are usually less significant than the hydraulic losses.

8.2.1 *Gear motors*

The gear motor of Fig. 8.4 is the simplest form of hydraulic motor with powers in the range to 20 kW. They are of fixed displacement and speed control can only be achieved by controlling the flow of hydraulic fluid to the motor. The internal leakage of fluid across the tips of the teeth, through the mesh and across the end faces of the gears make these the least efficient of the hydraulic motors.

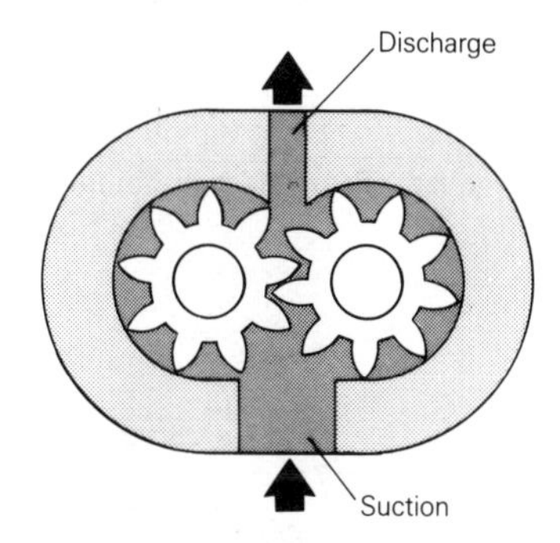

Fig. 8.4 Gear motor.

8.2.2 *Vane motors*

Vane motors have a higher speed and power rating than gear motors and are available with a variable displacement form as in Fig. 8.5, achieved by varying the eccentricity of the rotor.

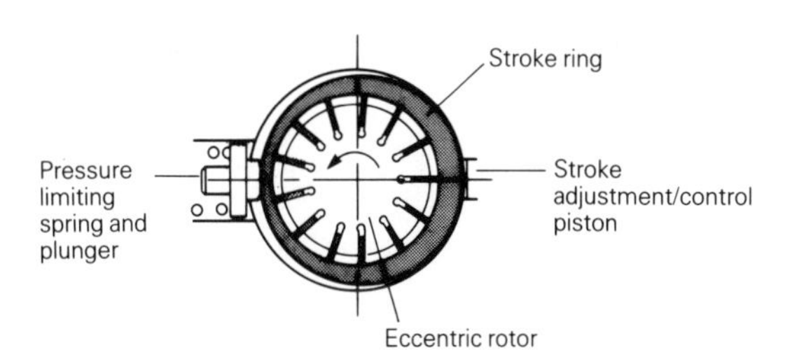

Fig. 8.5 Vane motor.

8.2.3 *Axial piston motors*

The two main varieties of axial piston motor are the bent axis machine of Fig. 8.6 and the variable displacement swash plate motor of Fig. 8.7. In the case of the former construction the specific displacement can be varied by adjusting the angle of the piston block relative to the drive shaft while in the case of the latter, variation of the swash plate angle has the same effect.

The torque produced by an axial piston motor is a function of the number of pistons and is subject to variations. This fluctuation, which may be troublesome in certain applications through the excitation of torsional vibrations, reduces as the number of pistons increases. For instance, for a swash plate motor with seven pistons the torque will vary in the range:

$$T_{max} = 2.18\, p \,.\, a \,.\, r \,.\, \tan\theta \qquad (8.4)$$

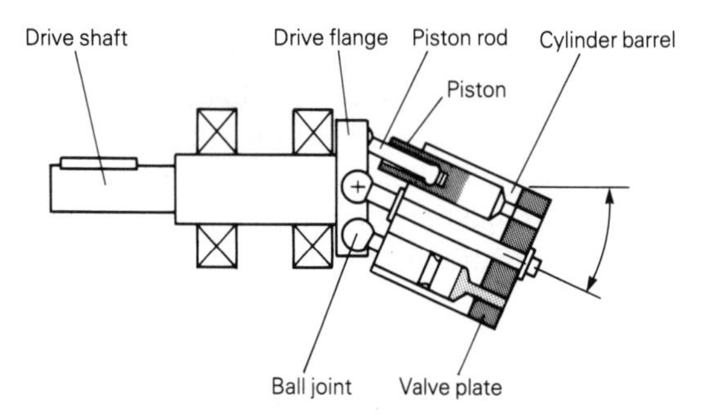

Fig. 8.6 Bent axis pump/motor.

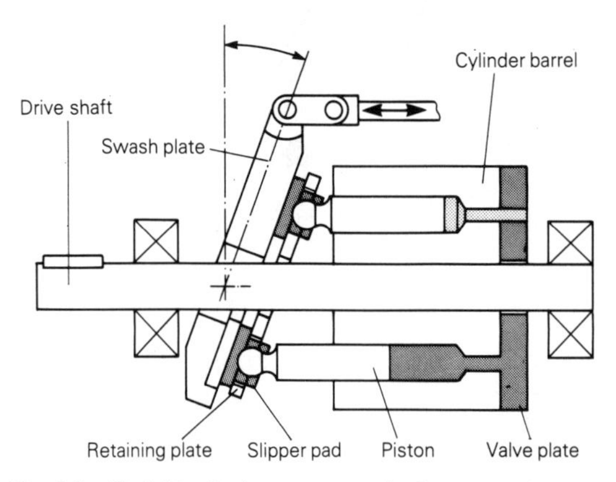

Fig. 8.7 Variable displacement swash plate pump/motor.

and

$$T_{\min} = 2.11\, p \,.\, a \,.\, r \,.\, \tan\theta \tag{8.5}$$

where:

a is the area of a piston;
p is the supply pressure;
r is the radius of the point of contact of the piston with the swash plate;
θ is the swash plate angle.

For control purposes the angle of the swash plate can be adjusted by means of an appropriate linear actuator.

8.2.4 *Radial piston motors*

Radial piston motors are generally constructed for low speed operation in the range to around 400 rev/min. Depending on their size, they are however capable of delivering very high torques with machines with outputs of 50 000 N m or more available. Figure 8.8 shows the construction of an inward working radial piston motor in which the motion of the pistons is transferred to the output shaft using sliding pads or connecting rods and an eccentric. The largest radial piston motors are of the cam ring type of Fig. 8.9.

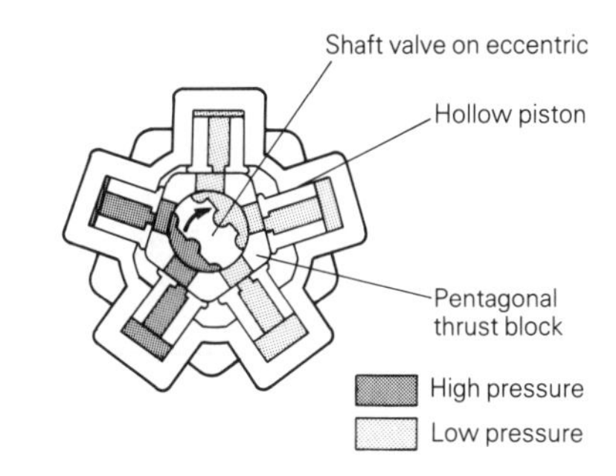

Fig. 8.8 Hydraulic motor, radial piston inward working type.

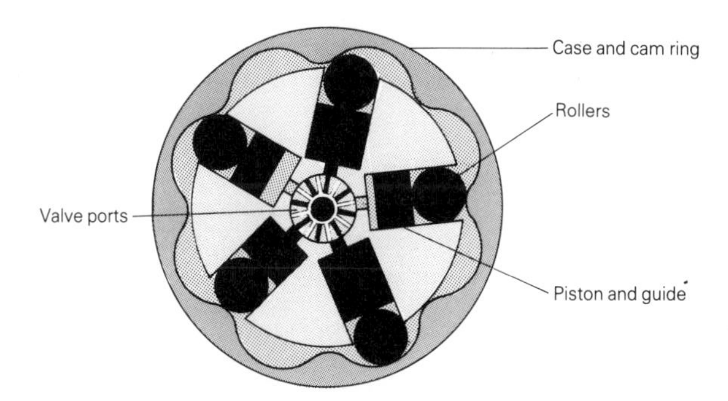

Fig. 8.9 Hydraulic motor, cam ring type.

Radial machines are usually of fixed displacement. However, with double banked versions of the inward working type or cam ring motors with an even number of cylinders it may be possible with the use of appropriate valving to disable half the cylinders, doubling the speed for the same flow rate.

8.3 Rotational hydraulic transmissions

The principle advantage of hydraulic motors is their high torque/weight ratio which for axial piston motors is of the order of 15 N m/kg. This is some ten times greater than for a typical induction motor based on the full load torque available at the shaft of the machine. Taken with the speed range available this means that a hydraulic motor can often be used to drive a mechanical system directly without the intervention of a reduction gearbox. This has significant advantages where there is a possibility of jamming of the driven load, for instance agitators and mixers.

In the case of an electrical drive with a reduction gearbox it is possible that up to 95% of the rotational kinetic energy of the entire system resides in the rotor of the electrical machine taking into account the individual inertias of the system components and in such circumstances the response of conventional trips may not be sufficient to prevent damage. Indeed, even after the supply has been disconnected the residual kinetic energy in the rotor of the machine may impart sufficient strain energy into the remainder of the system to cause a mechanical breakdown. Slipping couplings or fluid couplings may be used to provide additional protection but these add additional complexity to the system as well as occupying extra space.

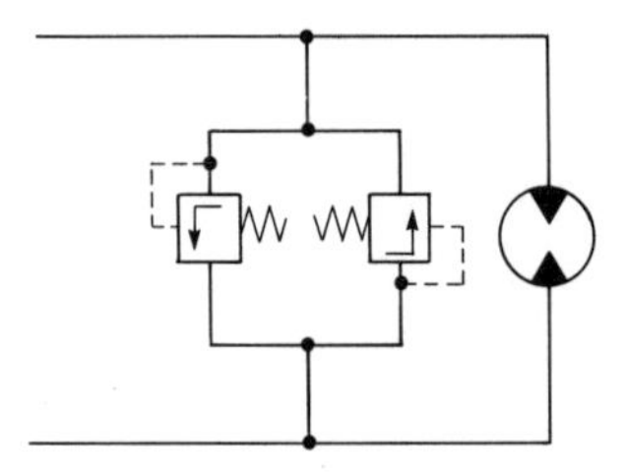

Fig. 8.10 Cross-line relief valves.

Protection against a normal overload is provided in the hydraulic system by system or cross-line relief valves as in Fig. 8.10.

The majority of hydraulic transmissions use a one pump per motor configuration as in Fig. 8.11. By using a variable capacity pump together with a variable capacity motor the generalized output characteristic of Fig. 8.12 is achieved.

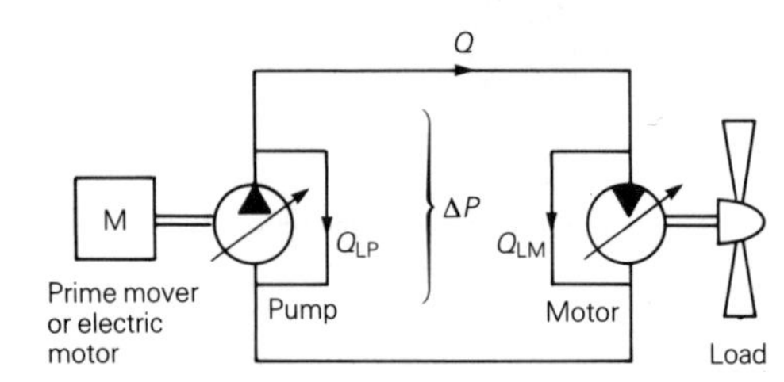

Fig. 8.11 Simple hydraulic transmission: Q_L is internal leakage flow.

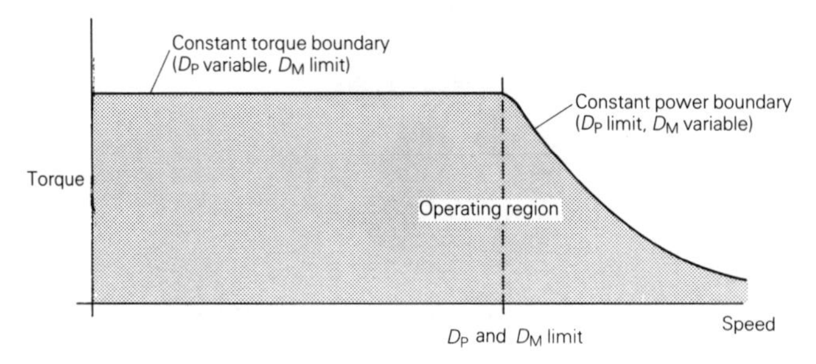

Fig. 8.12 Torque/speed characteristic for a combination of variable displacement pump and motor with pressure limitation.

Comment

Compare Fig. 8.12 with the combination of Figs 4.39 and 4.40 for the d.c. machine and 6.22 for the induction motor.

For any particular setting of the two machines, the speed and torque ratios are given by:

$$\frac{\omega_{\text{motor}}}{\omega_{\text{pump}}} = \frac{D_{\text{pump}}}{D_{\text{motor}}} = \frac{T_{\text{pump}}}{T_{\text{motor}}} \tag{8.6}$$

In selecting the control strategy for the hydraulic transmission the application must be taken into account. The main categories of application are:

- winching and reeling operations in which the torque is constant;
- applications such as centrifuges involving rapid accelerations in which $T\omega = \text{constant}$;
- situations such as a production lathe requiring a constant product of speed and force;
- hydrodynamic machine drives such as centrifugal pumps, fans and propeller type mixers in which $T = K\omega^2$.

8.3.1 *Hydraulic ring mains*

Conventionally, fixed displacement hydraulic motors are controlled from a constant pressure main by means of pressure compensated flow control valves as in Fig. 8.13. The disadvantage of this approach is seen from a consideration of the hypothetical power utilization diagram of Fig. 8.14. Particular adverse conditions exist where a

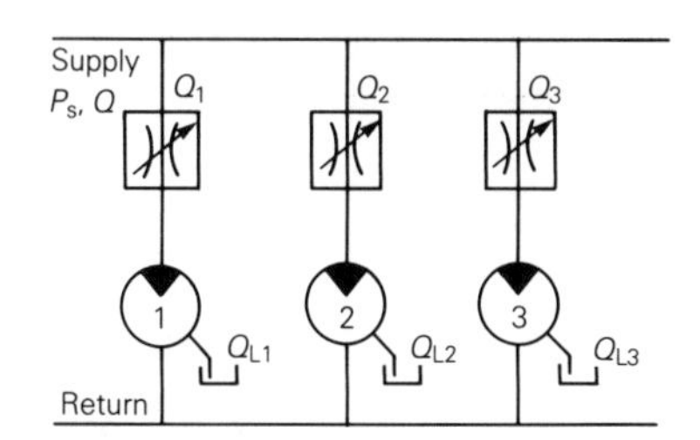

Fig. 8.13 Hydraulic ring circuit.

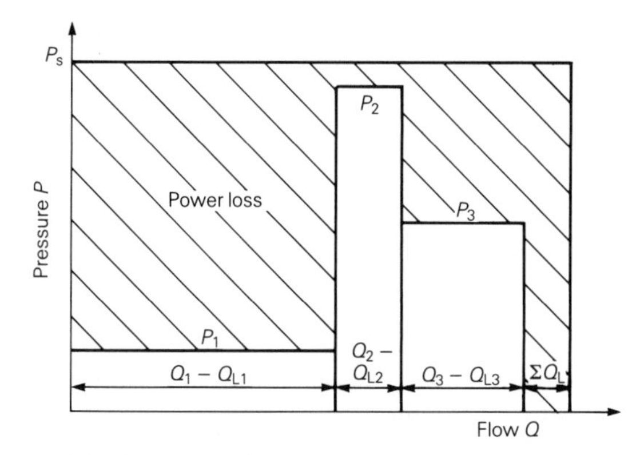

Fig. 8.14 Ring circuit power utilization diagram.

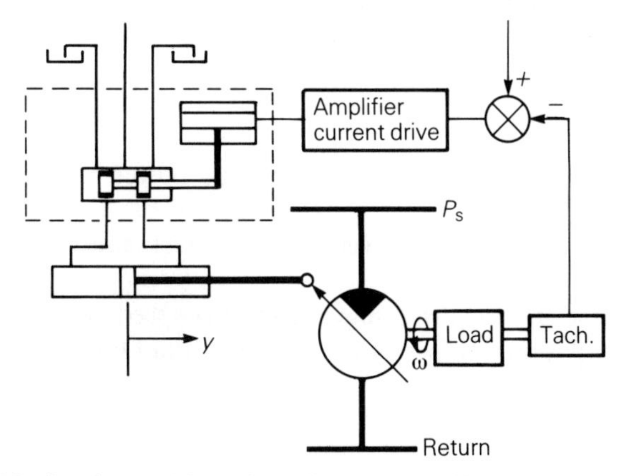

Fig. 8.15 Speed control by variable displacement with motor in ring circuit.

number of machines run at high speed but with low torques while others operating at higher torques set the system pressure. In these circumstances the hatched area of Fig. 8.14 represents the aggregate of the power lost in the flow control valves, plus a smaller loss resulting from the internal leakage flow.

The approach of Fig. 8.15 is one possible means of overcoming this limitation. Consider this system operating initially at point (a) on curve (i) of Fig. 8.16. If the system load characteristic now changes to that of curve (ii) on Fig. 8.16 – for instance as a result of a reduction of mixing viscosity – the speed of the motor would change from ω_1 to ω_2. In order to maintain the set point the speed of the motor would be reduced by reducing the displacement of the motor by varying the angle of the swash plate.

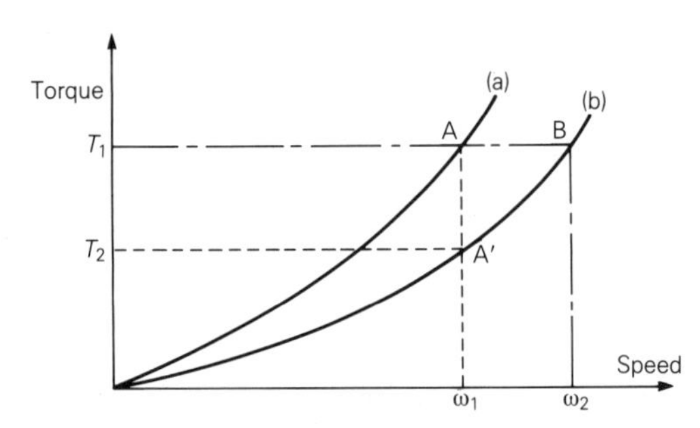

Fig. 8.16 Typical load characteristics.

8.4 Hydraulic motors – summary

The attributes of hydraulic transmissions can be summarized as follows.

- They provide high torques and powers with low weight and bulk.
- They are very resistant to mechanical shock.
- They are acceptable in areas with hazard classification.
- Low speed varieties have relatively low noise levels.
- They provide the ability to substantially remove the sources of noise (low speed, high torque motors only) and heat production from the working environment by using a remotely positioned pump and power pack with suitable insulation and heat removal facilities.

Disadvantages include the following.

- They require the provision of either a pump and power pack or a hydraulic ring main.
- Higher speed hydraulic motors (over 1000 rev/min) tend to generate high noise levels.
- Apart from the case of direct drive, high torque, low speed motors, system costs – for instance the installation of pipework – are higher than for electrical drives.

Appendix A Complex arithmetic

Phasors and complex numbers can be expressed in complex polar form in terms of their magnitude and phase angle. Consider the following.

$$a + jb = A \, . \, e^{j\theta} \qquad (A.1)$$

where A is the magnitude (or modulus) of the complex phasor or number such that

$$A = (a^2 + b^2)^{\frac{1}{2}} \qquad (A.2)$$

The phase angle θ is given by

$$\theta = \tan^{-1}\left(\frac{b}{a}\right) \qquad (A.3)$$

Equation (A.1) would more usually be written in phasor terms as:

$$a + jb = A\underline{|\theta} \qquad (A.4)$$

Comment

- The $(a + jb)$ form is referred to as the rectangular form of the equation.
- The $Ae^{j\theta}$ and $A\underline{|\theta}$ forms are referred to as the polar form of the equation.

A.1 Addition

Addition of a pair of complex numbers is carried out using the rectangular forms of the numbers and adding the real and complex terms separately. The result is a further complex number. Consider:

$$(3 + j) + (1 + 2j) = (3 + 1) + j(1 + 2) = 4 + 3j \qquad (A.5)$$

A.2 Subtraction

Subtraction of a pair of complex numbers is carried out using the rectangular forms of the numbers and subtracting the real and complex terms separately. The result is a further complex number. Consider:

$$(3 + j) - (1 + 2j) = (3 - 1) + j(1 - 2) = 2 - j \qquad (A.6)$$

A.3 Multiplication and division

There are no rules for the vector multiplication of phasors corresponding to those of vector and scalar product, however, complex phasors can be multiplied or divided by complex numbers to generate other phasors. Division of one complex phasor by another yields a complex number. In either case, the dimensions of the quantities involved must be taken into account.

A.3.1 *Multiplication*

Multiplication of a phasor by a complex number results in a new phasor whose magnitude is the product of the amplitude of the complex phasor and the modulus of the complex number at a phase angle which is the product of the individual phase angles. Consider:

$$(3+j)(1+2j)=3\times 1+1\times j+3\times 2j+j\times 2j=(3-2)+j(1+6)=1+7j \qquad (A.7)$$

Using polar form

$$(3+j)(1+2j)=3.16\underline{|18.43^\circ}\times 2.24\underline{|63.43^\circ}=7.08\underline{|81.86^\circ}=1+7j \qquad (A.8)$$

A.3.2 *Complex conjugate*

The complex conjugate of a complex number is obtained by changing the sign of the complex term in the rectangular form or the sign of the angle term in the polar form. The complex conjugate of a phasor is usually written as $\boldsymbol{A}^*$. For instance the complex conjugate of:

$$3+j=3.16\underline{|18.43^\circ}$$

is

$$3-j=3.16\underline{|-18.43^\circ}$$

The product of a complex number and its complex conjugate is then the (amplitude)^2 of the complex number. Consider:

$$(3+j)(3-j)=3.16\underline{|18.43^\circ}\times 3.16\underline{|-18.43^\circ}=10 \qquad (A.9)$$

A.3.3 *Division*

Division of a phasor by a complex number results in a new phasor whose magnitude is the divisor of the amplitude of the complex phasor and the modulus of the complex number at a phase angle which is the difference between the individual phase angles. Consider:

$$\frac{(3+j)}{(1+2j)}=\frac{(3+j)(1-2j)}{(1+2j)(1-2j)}=\frac{(5-5j)}{5}=1-j=1.41\underline{|-45^\circ} \qquad (A.10)$$

or

$$\frac{(3+j)}{(1+2j)}=\frac{3.16\underline{|18.43^\circ}}{2.24\underline{|63.43^\circ}}=1.41\underline{|-45^\circ} \qquad (A.11)$$

Appendix B Inertial loads

B.1 Tangentially driven loads

Tangentially driven loads include belts, pulleys, conveyors, rack and pinion drives and pinch wheels. For the arrangement of Fig. B.1, the torque required to accelerate the load at a constant angular velocity is given by:

$$T = J_1 + J_m + J_p \qquad \text{(B.1)}$$

in which

$$J_1 = M_1 . r^2 \qquad \text{(B.2)}$$

where M_1 is the mass of the load (kg), r is the radius of the pulley (m) and J_m and J_p are the inertias of the motor and the pulley, respectively (kg m^2 or N m/s).

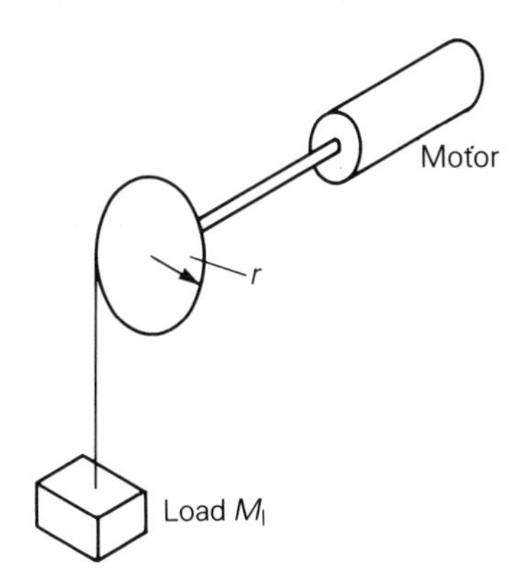

Fig. B.1 Tangentially driven load.

B.1.1 *Systems with a ratio change*

Consider the arrangements of Fig. B.2. For the belt drive of Fig. B.2(a), the effective inertia at the motor shaft is given by:

$$J_e = J_1 + J_2 . \left(\frac{D_1}{D_2}\right)^2 \qquad \text{(B.3)}$$

Similarly, for the geared drive of Fig. B.2(b):

$$J_e = J_1 + J_2 . \left(\frac{T_1}{T_2}\right)^2 \qquad \text{(B.4)}$$

where J_1 is the inertia of pulley 1 or gear 1, J_2 is the inertia at the load shaft of the combination of the load, the load shaft and pulley 2 or gear 2, D_1 and D_2 are the effective diameters of pulleys 1 and 2 and T_1 and T_2 are the number of teeth on gears 1 and 2.

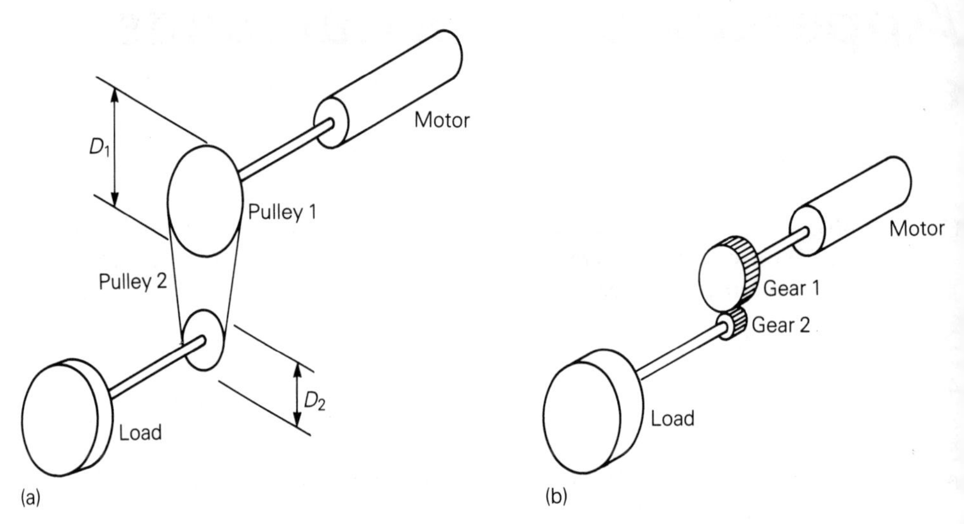

Fig. B.2 System with a ratio change: (a) belt drive; (b) geared drive.

B.2 Leadscrew driven loads

For the system of Fig. B.3, the torque required to accelerate the load at a constant linear acceleration a is given by:

$$T_a = \left(\frac{J_1}{e} + J_s + J_m\right)\frac{2\pi a}{p} \tag{B.5}$$

in which

$$J_1 = \frac{M_1 p^2}{(2\pi)^2} \tag{B.6}$$

where M_1 is the mass of the load (kg), p is the pitch of the leadscrew (m), e is the efficiency of the leadscrew and J_s and J_m are the inertias of the leadscrew and the motor (kg m^2 or N m/s).

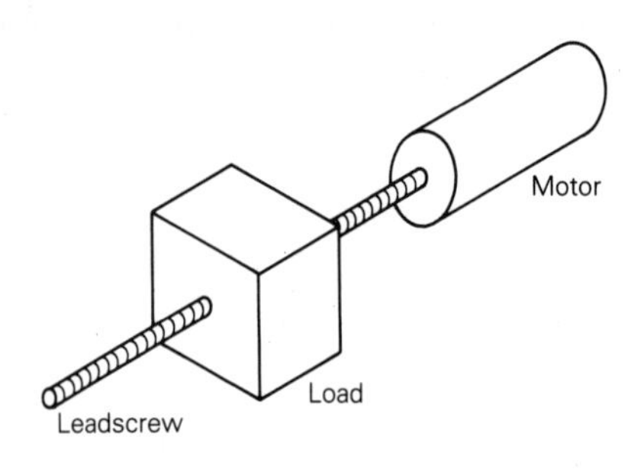

Fig. B.3 Leadscrew driven load.

Appendix C Answers to exercises

Chapter 1

1 $I=23.97\angle{-14.34^\circ}$ A; power $=5574$ W; power factor $=0.969$ lagging
2 39.19 kVAR
3 power factor $=0.972$ leading; $S_s=55.55$ kVA; $I=8.75\angle{13.59^\circ}$ A
4 $I_1=29\angle{-45.25^\circ}$ A; I_2 and I_3 are equal in magnitude and displaced by -120° and -240°, respectively. The phase impedance is $(4.22+1.15j)=4.37\angle{15.24^\circ}$ Ω
5 $I_{pu}=1.147\angle{-25.33^\circ}$ pu or $I=60.2\angle{-25.33^\circ}$ A
6 (a) $I_1=33\angle{-43.8^\circ}$ A; $I_2=40.8\angle{-172.73^\circ}$ A; $I_3=32.6\angle{59.13^\circ}$ A
(b) $I_1=23.9\angle{-30^\circ}$ A; $I_2=33.9\angle{-172.73^\circ}$ A; $I_3=321.4\angle{59.13^\circ}$ A; $I_n=16.1\angle{98.65^\circ}$ A
7 $I_1=6.86\angle{-30^\circ}$ A; $I_2=4.75\angle{-166.38^\circ}$ A; $I_3=4.74\angle{106.18^\circ}$ A; power $=920.9$ W

Chapter 2

1 When the transistor is off, effective voltage is 12 V, therefore load line intercepts the V_g axis at 12 V. With the transistor on the intercept of the load line with the I_g axis determines R. The load line must avoid the gate temperature limits and the gate power limit.
2 $\omega L=151$ Ω which is $\gg$ load resistance, hence the d.c. current is effectively constant. (a) $V_0=187.1$ V; $I_{dc}=51.97$ A and is both the average and the r.m.s. current in the absence of ripple. (b) Each thyristor conducts for half a cycle, therefore $I_{tave}=25.99$ A and $I_{trms}=36.75$ A
3 (a) $I_{dc}=55.98$ A. (b) Each thyristor conducts over an angle of $(\pi-\alpha)$, when $I_{tave}=23.3$ A and $I_{trms}=36.1$ A
4 $\beta=63.5^\circ$; mean thyristor current $=10$ A; r.m.s. thyristor current $=17.32$ A
5 The limit is set by average current, hence $I_{dc}=75$ A. Maximum power occurs with $\alpha=0^\circ$, when $V_0=970.7$ V and $P_{max}=72.8$ kW. Average power dissipated in the thyristor is 32 W when the thermal resistance of the heat sink is 1.965 °C/W
6 (a) power factor $=0.92$, $\mu=0.953$ and $\cos\phi_1=0.966$; (b) power factor $=0.74$, $\mu=0.854$ and $\cos\phi_1=0.866$
7 $t_{on}=0.3$ ms; $I_L=7.2$ A; $\Delta I_r=0.24$ A; $I_{r;rms}=0.693$ A
8 (a) $I_{load}=16.33$ A, $I_{thyristor}=11.5$ A, power $=12$ kW; (b) $I_{load}=18.85$ A, $I_{thyristor}=13.33$ A, power $=15.99$ kW

Chapter 3

1 $\Phi=2.8\times10^{-5}$ Wb
2 1.61 A
3 $L_m=2.51$ mH; $R_c=368$ W
4 $Hl=4767$ ampere-turns

5 Need to increase effective areas of sections (a) and (b) to reduce the flux density to around 1.4 T when for both sections cross-sectional area = 820 cm^2 when $\boldsymbol{H}l$ = 5740 ampere-turns
6 $\boldsymbol{B}$ = 1.11 T
7 $V_1 = 204\angle 1.78°$; $I_1 = 20.4\angle -1.7°$; efficiency = 96.3%
8 (a) $V_2 = 1860\angle -3.30°$ V; (b) $I_1 = 25.97\angle -40.8°$ A; (c) power in = 3932 W; (d) power out = 3732 W
9 415 V

Chapter 4

1 (a) 883 rev/min; (b) 994 rev/min
2 (a) $\omega_2/\omega_1 = 0.99$, i.e. a 1% reduction in speed; (b) $\omega_2/\omega_1 = 0.975$, i.e. a 2.5% reduction in speed
3 (a) $I_a = 226.3$ A, steady state speed = 1460 rev/min; (b) $I_a = 338.6$ A, steady state speed = 1060 rev/min; (c) $I_a = 320$ A, steady state speed = 908 rev/min
4 2009 rev/min
5 (a) $V = 96.3$ V; (b) $I_f = 2.28$ A
6 $I_{a,max} \sim 30$ A
7 A two-step starter is required with steps of 0.3 Ω and 0.15 Ω

Chapter 5

1 94.8%
2 (a) $I_a = 26.25\angle 0°$ A, $E_a = 1270\angle 1.93°$ V; (b) $I_a = 32.8\angle 36.87°$ A, $E_a = 1240\angle 2°$ V; (c) $I_a = 30.9\angle -31.79°$ A, $E_a = 1297\angle 1.9°$ V
3 $\delta = \angle -11.37°$, $E_a = 274.6\angle -11.37°$ V
4 power factor = 0.73 lagging; $\delta = \angle 18.5°$
5 (a) $I_t = 89\angle 32.33°$ A; (b) $E_a = 572.5\angle 19.28°$ V (phase)
6 The machine remains in synchronism

Chapter 6

1 (a) $T_{max} = 322$ N m; (b) $T = 230$ N m; (c) $T = 126.5$ N m; (d) $R_{add} = 3.66$ Ω/phase; (e) $R_{add} = 1.21$ Ω/phase; (f) 1014 rev/min; (g) $T_{max} = 478$ N m; (h) $T = 288.5$ N m; (i) 16 150 W
2 $T = 69.7$ N m
3 $I_r^2 R_r = I_r'^2 R_r'$ = rotor copper loss = 295 W; $P_{ag} = 5535$ W
4 On starting the voltage/phase is $415/\sqrt{3}$ V and the starting current is therefore reduced by $\frac{1}{3}$ from the short circuit current. Hence the ratio of starting current to full load current is $\frac{55}{32.7} = 1.68$
5 (a) $s = -0.0684$, speed = 1068.4 rev/min; (b) gearbox ratio = 71.2:1; (c) windspeed = 4.9 km/h
6 Power = 1064 W; torque = 7.2 N m
7 $V_f = 99.5\angle -0.13°$ V; $V_b = 6.78\angle 0.42°$ V

Further reading

Bird, B.M., King, K.G. and Pedder, D.A.G., 1993, *An Introduction to Power Electronics*, John Wiley.

Bradley, D.A., 1987, *Power Electronics*, Van Nostrand Reinhold.

Bradley, D.A., Dawson, D., Burd, N.C. and Loader, A.J., 1991, *Mechatronics – Electronics in Products and Processes*, Chapman & Hall.

Del Toro, V., 1985, *Electric Machines*, Prentice-Hall.

Dewan, S.G., Straughen, A.R. and Slemon, G.R., 1984, *Power Semiconductor Drives*, Wiley.

Fitzgerald, A.E., Kingsley, C. and Umans, S.D., 1984, *Electric Machinery*, McGraw-Hill.

Gray, C.B., 1989, *Electrical Machines and Drive Systems*, Longman.

Hindmarsh, J., 1980, *Worked Examples on Electrical Machines and Drives*, Pergamon.

Hindmarsh, J., 1984, *Electrical Machines and their Applications*, Pergamon.

Kenjo, T., 1984, *Stepping Motors and their Microprocessor Control*, Oxford Scientific.

Lander, W.C., 1993, *Power Electronics*, 3rd edn, McGraw-Hill.

McLaren, P.G., 1984, *Elementary Electric Power and Machines*, Ellis Horwood.

O'Kelly, D., 1991, *Performance and Control of Electrical Machines*, McGraw-Hill.

Sarma, M.S., 1985, *Electric Machines; Steady State Theory and Dynamic Performance*, West.

Shepherd, W. and Hulley, L.N., 1987, *Power Electronics and Motor Control*, Cambridge University Press.

Slemon, G.R. and Straughen, A.R., 1980, *Electric Machines*, Addison-Wesley.

Index